ANIMAL AND PLANT
Anatomy

VOLUME CONSULTANTS

• Allan Bornstein, *Southeast Missouri State University, MO* • John Gittleman, *University of Virginia, VA*
• Tom Jenner, *Academia Británica Cuscatleca, El Salvador* • Alan Leonard, *Florida Institute of Technology, Melbourne, FL* • Kieran Pitts, *Bristol University, England* •
Erik Terdal, *Northeastern State University, Broken Arrow, CA*

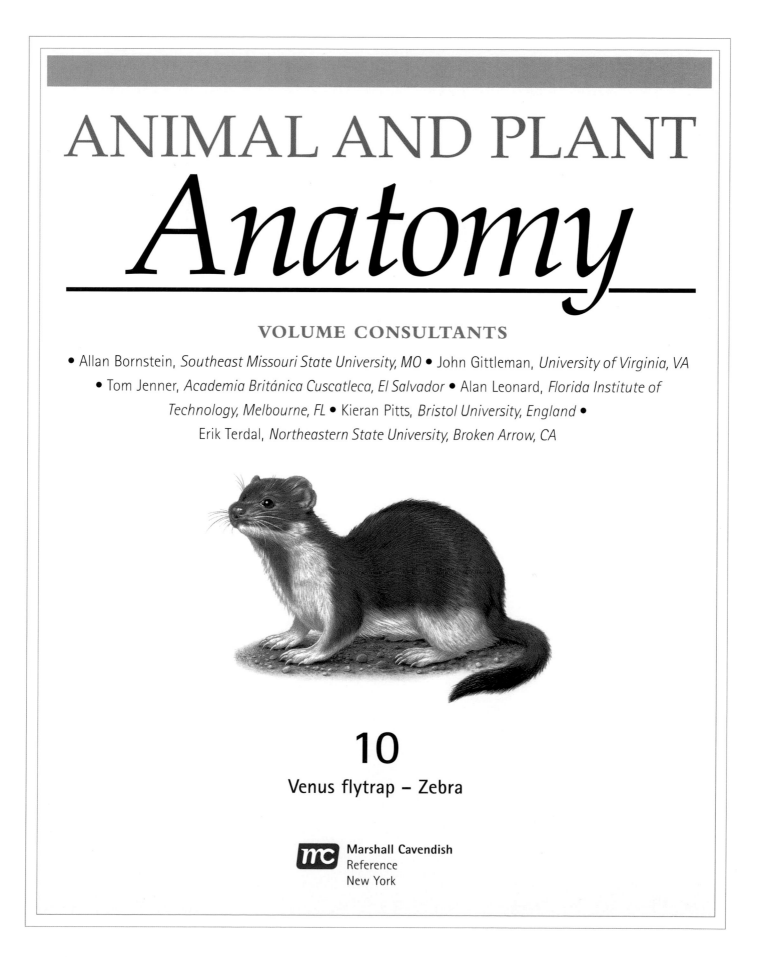

10
Venus flytrap – Zebra

mc **Marshall Cavendish**
Reference
New York

CONTRIBUTORS

Roger Avery; Richard Beatty; Amy-Jane Beer; Erica Bower; Trevor Day; Erin Dolan; Bridget Giles; Natalie Goldstein; Tim Harris; Christer Hogstrand; Rob Houston; John Jackson; Tom Jackson; James Martin; Chris Mattison; Katie Parsons; Ray Perrins; Kieran Pitts; Adrian Seymour; Steven Swaby; John Woodward.

CONSULTANTS

Barbara Abraham, Hampton University, VA; Glen Alm, University of Guelph, Ontario, Canada; Roger Avery, Bristol University, England; Amy-Jane Beer, University of London, England; Deborah Bodolus, East Stroudsburg University, PA; Allan Bornstein, Southeast Missouri State University, MO; Erica Bower, University of London, England; John Cline, University of Guelph, Ontario, Canada; Trevor Day, University of Bath, England; John Friel, Cornell University, NY; Valerius Geist, University of Calgary, Alberta, Canada; John Gittleman, University of Virginia, VA; Tom Jenner, Academia Británica Cuscatleca, El Salvador; Bill Kleindl, University of Washington, Seattle, WA; Thomas Kunz, Boston University, MA; Alan Leonard, Florida Institute of Technology, FL; Sally-Anne Mahoney, Bristol University, England; Chris Mattison; Andrew Methven, Eastern Illinois University, IL; Graham Mitchell, King's College, London, England; Richard Mooi, California Academy of Sciences, San Francisco, CA; Ray Perrins, Bristol University, England; Kieran Pitts, Bristol University, England; Adrian Seymour, Bristol University, England; David Spooner, University of Wisconsin, WI; John Stewart, Natural History Museum, London, England; Erik Terdal, Northeastern State University, Broken Arrow, OK; Phil Whitfield, King's College, University of London, England.

Marshall Cavendish

99 White Plains Road
Tarrytown, NY 10591–9001

www.marshallcavendish.us

© 2007 Marshall Cavendish Corporation

Library of Congress Cataloging-in-Publication Data

Animal and plant anatomy.

    p. cm.

    ISBN-13: 978-0-7614-7662-7 (set: alk. paper)

    ISBN-10: 0-7614-7662-8 (set: alk. paper)

    ISBN-13: 978-0-7614-7674-0 (vol. 10)

    ISBN-10: 0-7614-7674-1 (vol. 10)

    1. Anatomy. 2. Plant anatomy. I. Marshall Cavendish Corporation. II.
Title.

QL805.A55 2006
571.3--dc22

                    2005053193

Printed in China
09 08 07 06 1 2 3 4 5

MARSHALL CAVENDISH

Editor: Joyce Tavolacci
Editorial Director: Paul Bernabeo
Production Manager: Mike Esposito

THE BROWN REFERENCE GROUP PLC

Project Editor: Tim Harris
Deputy Editor: Paul Thompson
Subeditors: Jolyon Goddard, Amy-Jane Beer, Susan Watts
Designers: Bob Burroughs, Stefan Morris
Picture Researchers: Susy Forbes, Laila Torsun
Indexer: Kay Ollerenshaw
Illustrators: The Art Agency, Mick Loates, Michael Woods
Managing Editor: Bridget Giles

Contents

Venus flytrap

KINGDOM: Plantae ORDER: Nepenthales
FAMILY: Droseraceae GENUS: *Dionaea*

The Venus flytrap is a carnivorous plant. Carnivorous plants catch insects and other small animals to supplement their diet. The traps of the Venus flytrap, like those of all carnivorous plants, are modified leaves. The leaves snap shut when an insect knocks against trigger hairs. The victim is then squashed and bathed in digestive juices. The closing of the trap is one of the fastest movements in the plant kingdom.

Anatomy and taxonomy

There are more than 600 species of carnivorous plants, from 18 genera, spread across 8 families. They occur on every continent except Antarctica. Most carnivorous plants live in swamps or bogs, where the soil has few nutrients. The plants capture insects and other small animals to supplement their diet.

● **Plants** One of the key characteristics of most plants is that they are green. The color comes from a pigment called chlorophyll. This pigment allows plants to create their own food from just air and water, using energy from sunlight in a process called photosynthesis. The other major groups of multicellular organisms (animals and fungi) cannot make their own food and depend on plants for their existence.

● **Seed-bearing plants** There are two major groups of seed plants. Gymnosperms (meaning "naked seeds") include pine trees and other conifers. Angiosperms (meaning "enclosed seeds") are the flowering plants and have their ovules inside an ovary, which ripens into a fruit containing the seeds.

● **Flowering plants** Flowering plants, or angiosperms, are divided into the monocotyledons (monocots) and dicotyledons (dicots). Monocots have one cotyledon, or seed leaf, within the seed—hence the group's name. Monocots usually have long, narrow leaves with parallel veins and flower parts grouped in threes. Monocots include palms, lilies, grasses, and orchids.

● **Dicots** The broad-leaved plants, or dicots, have leaves with netlike veining and are very variable in shape. The seedlings usually have two cotyledons. The flowers tend to have parts grouped in fours or fives. Most of our familiar plants are dicots, including roses and apple trees. All the carnivorous plants are dicots. They fall into two separate groups: one with flowers with bilateral symmetry (having only one plane of symmetry) and the other with flowers with radial symmetry (cutting through the center of the flower in any direction produces two halves that are mirror images of each other).

Carnivorous plants with bilateral symmetry are organized within the family Lentibulariaceae. All members of this family of plants have bilateral symmetry

▼ *Carnivory is relatively widespread among dicots, occurring in at least eight families from three different orders. This suggests that carnivory has evolved more than once in plants.*

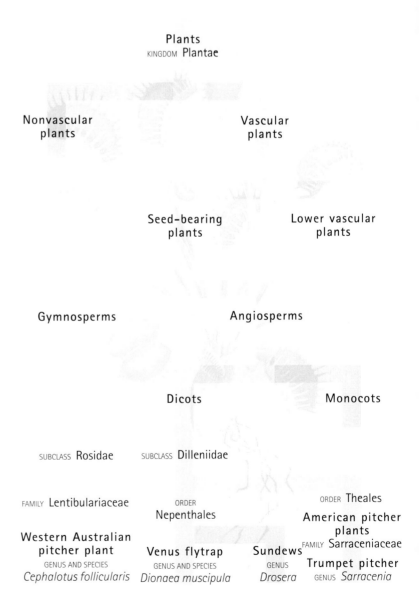

Plants
KINGDOM Plantae

Nonvascular plants — Vascular plants

Seed-bearing plants — Lower vascular plants

Gymnosperms — Angiosperms

Dicots — Monocots

SUBCLASS Rosidae — SUBCLASS Dilleniidae

FAMILY Lentibulariaceae — ORDER Nepenthales — ORDER Theales

American pitcher plants

Western Australian pitcher plant — Venus flytrap — Sundews — FAMILY Sarraceniaceae

GENUS AND SPECIES *Cephalotus follicularis* — GENUS AND SPECIES *Dionaea muscipula* — GENUS *Drosera* — Trumpet pitcher GENUS *Sarracenia*

and are carnivorous. They include butterworts and bladderworts. Most of the other carnivorous plants are in the subclass Dilleniidae, with the exception being the pitcher plants from south Western Australia, which are in the subclass Rosidae.

● **Dilleniidae** This is a large, diverse class with three orders that contain carnivorous plants. The rainbow plant is a glandular, sticky plant from Australia, in the order Ericales. The American pitcher plants are in the order Theales, family Sarraceniaceae. They consist of the cobra lily—not a true lily, because lilies are monocots— the trumpet pitchers, and the South American genus of primitive swamp pitchers *Heliamphora*. Other carnivorous plants in the subclass Dilleniidae are in the order Nepenthales.

● **Nepenthales** The families of plants within the order Nepenthales are carnivorous. They include Dioncophyllaceae, Nepenthaceae, and Droseraceae. *Triphyophyllum peltatum* (in the family Dioncophyllaceae) is an unusual woody climbing plant from West Africa. It produces three different types of leaves. The first are regular leaves, produced on a juvenile shoot that grows to about 3 feet (1 m) high. Then, a second type of leaf grows, which is very long and thin, with glandular hairs that are sticky and trap insects. After a year or two, the plant produces a climbing stem that has small nonglandular leaves with hooked tips. These leaves help the liana (vine) climb to the treetops.

The tropical pitcher plants, family Nepenthaceae, are climbing plants found in Asia. Leaves form a "jug," or

▲ *With leaves modified into traps, the Venus flytrap is one of the most distinctive plants in the natural world.*

pitcher, that grows from a tendril at the end of a long leaf. The largest pitchers grow on *Nepenthes rajah*. These pitchers reach 14 inches (35 cm) long and 7 inches (18 cm) across and can hold 0.4 gallon (2 l) of fluid.

● **Droseraceae** The sundew family has about 110 species. These plants occur throughout the world, especially in wet places. Most classifications include four genera: *Aldrovanda*, *Drosera*, *Drosophyllum*, and *Dionaea*. However, genetic studies suggest that *Drosophyllum*, which has only one species, the Portuguese sundew, is more closely related to the genus *Triphyophyllum*, which is a member of the family Dioncophyllaceae in the order Nepenthales.

Sundews (genus *Drosera*) have leaves covered with sticky "tentacles" that can bend toward trapped prey. The movements are slow but perceptible. The Portuguese sundew is also covered with sticky hairs, but unlike other sundews, it has hairs that do not move. The waterwheel plant, which is the only species in the genus *Aldrovanda*, is a rootless, submerged aquatic plant from Europe, Asia, Africa, and Australia. It has whorls of leaves, each with a trap similar to a tiny version of that of the Venus flytrap, with trigger hairs and two lobes that snap together.

● **Genus *Dionaea*** This genus has only one species—the Venus flytrap, which is named for Venus, the Roman goddess of love. The Venus flytrap grows wild in a narrow strip of boggy coastal land about 10 miles (16 km) wide and 100 miles (160 km) long, straddling North Carolina and South Carolina.

EXTERNAL ANATOMY The Venus flytrap has leaves that are highly modified into deadly insect traps. There are three hairs on each lobe that, if touched, trigger the leaf to snap shut. *See pages 1302–1305.*

INTERNAL ANATOMY Vascular tissues transport water, nutrients, and the products of photosynthesis and digestion around the plant. Photosynthesis occurs in organelles called chloroplasts inside leaf cells. *See pages 1306–1307.*

DIGESTIVE SYSTEM Glands secrete digestive fluid that dissolves the soft parts of insects. *See pages 1308–1311.*

REPRODUCTIVE SYSTEM The small white flowers are pollinated by insects. The flowers are held on long stalks that keep pollinators away from the dangerous leaves below. *See pages 1312–1313.*

External anatomy

COMPARE the leaves of the Venus flytrap with those of the *SAGUARO CACTUS*. Both types of leaves are highly modified. The leaves of the Venus flytrap have evolved to catch insect prey. In contrast, the leaves of the saguaro are reduced to spines that minimize water loss and protect the plant from herbivores.

The Venus flytrap is similar to many other small herbs in its basic structure. Its incredible traps are modified leaves. The plant does not have aerial (vertical) stems, and all the leaves are arranged in a rosette, forming a circular pattern around the growing point in the center. When the plant flowers, a tall flower stalk emerges from the center.

Roots and rhizomes

Roots anchor the plant and absorb water and nutrients from the soil. The roots of the Venus flytrap are short and fleshy, and they probably also have a storage function. Most roots are within the top 4 inches (10 cm) of the soil, though a few can grow to a depth of 1 foot (30 cm). The Venus flytrap spreads over an area of soil by growing rhizomes. These are swollen stems that run horizontally just below the surface of the soil.

▼ *The northern pitcher plant grows in acidic bogs in North America. Its pitchers act as pitfall traps, containing fluid that digests insects. Unlike the Venus flytrap, pitcher plants trap prey passively.*

CLOSE-UP

Leaf glands

There are two types of glands on a Venus flytrap leaf, both just visible to the unaided eye. Digestive glands in the center of the lobes are dark red (if the plant has grown in good sunlight). Around the rim of the leaf, near the teeth, are nectar glands that produce a sugary liquid attractive to insects. This narrow band of nectar allows smaller insects to feed without triggering the trap. However, larger insects that could make a worthwhile meal are forced to position themselves closer to the center of the trap, where they risk touching the trigger hairs.

Because rhizomes are stems rather than roots, they have buds, from which new plants sprout. (Roots cannot produce buds.) With new plants sprouting at varying distances from the parent plant, an area soon becomes a tangle of deadly leaves in which any insect or spider is in danger wherever it walks.

Deadly leaves

All the leaves of a Venus flytrap are modified into highly efficient traps. However, they are still green and therefore able to photosynthesize. The leaf stalk, or petiole, provides a surface for photosynthesis. Enlarged leaf stalks are called phyllodes. They are thick and fleshy, serving as a food storage area.

At the end of the petiole is the leaf blade, or lamina. This is the part of the Venus flytrap that is modified into the trap. It has two kidney-shape lobes, fringed with 14 to 20 sharp projections, which are held at a slight angle to the lamina. The teeth of the two lobes interlock when the trap closes. The lobes are normally held at an angle of 40 to 50 degrees. On the upper surface of each lobe are three sensitive hairs arranged in a triangle. These structures are the trigger hairs. If an insect bends them, they trigger the trap to snap shut.

Flypaper, pitfalls, snap traps, and suction

Plants have devised many intricate structures with which they capture animals. All of the structures are modified leaves. The most simple are the passive, sticky "flypaper traps" of the butterworts and the rainbow plant. Butterworts have rosettes of flat, sticky leaves. Any insects that walk over them become glued to the surface and eventually die. The edges of the leaf gradually curl inward, and the insect is digested.

▼ *The Venus flytrap does not have aerial stems. Instead, the unusual leaves attach directly to the roots.*

Each lobe has 14 to 20 **projections**. *The projections of the two lobes interlock when the trap closes.*

The kidney-shape **lobes** *are modified leaf blades. They are green on the outside and red inside.*

The winged leaf stalk, or petiole, is thick and fleshy and is called a **phyllode**. *It is green and therefore photosynthesizes.*

Roots grow from the base of the leaf stems.

Short, fleshy **roots** *anchor the plant and absorb water and nutrients from the soil.*

6 inches
(15 cm)

4 inches
(10 cm)

▲ Nepenthes kinabaluensis *is a pitcher plant that grows at high altitudes on Mount Kinabalu in Sabah, Borneo. It is a natural hybrid of two other pitcher plants:* N. rajah *and* N. villosa.

The sundews also have sticky leaves. A droplet of sticky glue is produced from a gland at the tip of a stalk, which is called a tentacle. The struggling insect triggers a movement response in nearby tentacles, which bend toward the insect, covering the animal in even more fluid. In some species of sundews—and depending on the location of the insect—the end of the leaf can even roll over, surrounding the victim.

GENETICS

From leaf to pitcher

The swamp pitcher has relatively simple pitchers that can be easily envisaged as normal leaves that have curled around and joined at their base. Occasionally, unrelated plants, such as *Codiaeum variegatum*, grow pitcher-type leaves. These deformities are the result of a single gene mutation.

Thieves or partners?

Insects called capsid bugs live on many sundews. The bugs are able somehow to walk on the sticky leaf surface without being caught. They eat the trapped insects, but although they are "stealing" from the plant, the sundew may gain more benefit from this relationship than first appears. The bugs defecate on the plant, so the plant still receives much of the prey's nutrients, but in a recycled form.

This process is relatively slow. The bending of the tentacles is visible, but nowhere near as fast as Venus flytrap closure. The sundews rely on the stickiness of the fluid to hold the victim.

Pitcher plants present a different hazard. These plants have passive pitfall traps that do not depend on any plant movement to catch insects. The pitcher is a leaf modified into a hollow container that holds liquid. The plant attracts insects using scent, color, and patterns. When insects arrive on the leaf, they find nectar, usually around the rim of the pitcher. Feeding on the nectar is hazardous, because just below it is a very slippery surface. When an insect loses its foothold and falls inside, slippery walls and downward-pointing hairs prevent it from crawling out. It drowns in the fluid at the bottom and is slowly digested and absorbed by the plant.

There are various types of pitcher plants. Swamp pitchers are the least complex of the group, with simple, open vessels that fill with rainwater. The tropical pitcher plants, in the genus *Nepenthes*, are more complex. Most are climbing plants, and their pitchers form on the end of leaves that hang down then curl up at the end to form the "jug." The jugs usually have a lid that in historical texts was described as closing to trap insects, but that is not the case.

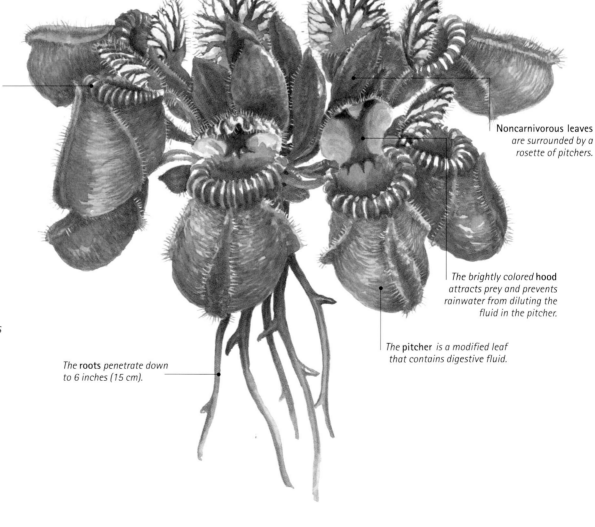

The pitchers have a teethlike rim, *which produces nectar to attract insects and spiders that slip into the pitcher and drown.*

Noncarnivorous leaves *are surrounded by a rosette of pitchers.*

The brightly colored **hood** *attracts prey and prevents rainwater from diluting the fluid in the pitcher.*

The roots *penetrate down to 6 inches (15 cm).*

The **pitcher** *is a modified leaf that contains digestive fluid.*

▶ **Australian pitcher plant**
This ground-hugging carnivorous plant grows wild in coastal regions of southwestern Australia. It has clumps of thumb-size pitchers and backward-curving teeth inside the pitcher that prevent unwitting victims from climbing out.

The lid acts more as a flag to attract prey and as an umbrella to keep rainwater from diluting the digestive fluids.

American pitcher plants are not climbing plants; they form low-growing rosettes. The cobra lily has a very hooded pitcher with a forked appendage, hanging over the pitcher entrance, that looks a little like a snake's tongue. The appendages attract insects and are covered with nectar. The hood of the pitcher has numerous "windows": transparent patches of tissue through which sunlight can pass. These windows are thought to misdirect any insects under the hood inside the pitcher. If you have ever tried to guide flies out of a window, you will know how they tend to fly toward the light and are reluctant to pass the shadow of a window frame, even if freedom lies on the other side. It is highly likely that the windows of hooded pitchers act in the same way. Even though the exit is close by, flies will not fly downward, away from the light. Eventually, the insects become so exhausted that they drop into the liquid and perish.

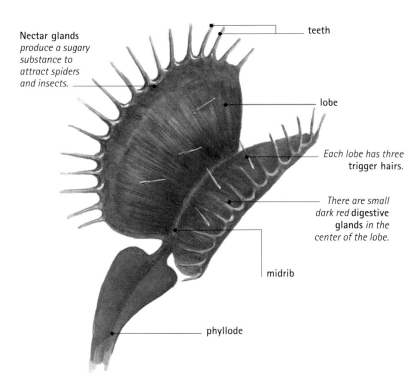

Nectar glands *produce a sugary substance to attract spiders and insects.*

teeth

lobe

Each lobe has three trigger hairs.

There are small dark red digestive glands *in the center of the lobe.*

midrib

phyllode

▲ Venus flytrap
Each lobe of the trap has three trigger hairs, as well as nectar-secreting glands that attract insects and spiders, and digestive glands that release juices to break down the soft parts of the prey.

EVOLUTION

Evolution of carnivory

To be classed as truly carnivorous, a plant must be able to attract prey, capture and retain it, digest it, absorb the digested material, and be shown to derive some benefit from it. Although carnivory appears to be a bizarre way of life for a plant, it can theoretically evolve relatively easily. Because nitrogen is usually in short supply in the soil, many plants probably carry out primitive carnivory (protocarnivory) to some extent. A simple form is "tank" protocarnivory, which occurs in bromeliads and some epiphytic orchids. The bases of bromeliad leaves form a tank that catches water. Litter and dead insects collect in this tank, and their nutrients are absorbed as they decay. Pitfall traps are likely to have evolved over millions of years from this type of protocarnivory; the simple swamp pitchers are the closest example.

Carnivorous plants that use "flypaper" traps probably evolved from glandular or "sticky" protocarnivores. Many plants are sticky as a defense against small herbivorous insects. Those insects that venture onto the plant stick to it and die there. Research has shown that even innocuous plants like the sticky purple geranium have carnivorous tendencies because they produce protease enzymes that can break down dead insects. It is not known how the complex traps of the Venus flytrap evolved, but the reaction is very similar to that of the sensitive plant *Mimosa*, which folds its leaves in defense when touched.

Venus flytrap is the most familiar snap trap, but the closely related waterwheel plant moves even faster. This rootless, aquatic plant floats just below the water's surface. It has whorls of eight or so traps around the stem. Each trap is similar to that of the Venus flytrap but is tiny, only 0.08 inch (2 mm) across. The trap closes with amazing speed, around ¹⁄₁₀₀th of a second; that is one of the most rapid movements in plants.

Bladderworts have one of the most complex trapping mechanisms in the plant kingdom. Most species are aquatic and live underwater. The traps are tiny bladders on their feathery leaves. Each bladder is armed with three trigger hairs and has a trapdoor that opens inward. The pressure inside the bladder is kept low because the plant actively pumps water out through the walls. This pulls in the sides of the bladder. If a water flea or other small water animal swims past and touches the trigger hairs, the trapdoor snaps open and the animal is sucked inside. Enzymes from the glands in the bladder digest the animal.

Internal anatomy

COMPARE the leaf cross section of the Venus flytrap with that of a *POTATO*. Potatoes and most other dicots have a layer of tall, thin, tightly packed cells called palisade mesophyll, where most of the photosynthesis takes place. In contrast, the Venus flytrap has no palisade cells, presumably because these cells would interfere with the leaf-closing mechanism.

Despite its bizarre mode of feeding, the key elements of the Venus flytrap's internal structure are the same as those for other plants. The structure consists of tissues that support the plant and provide its shape, and tissues that carry out photosynthesis, nutrient absorption, and transport. The vascular tissues transport water, nutrients, and the products of photosynthesis around the plant and between the roots and leaves. There are two main components of vascular tissue: xylem and phloem. Xylem cells conduct water and minerals from the roots to the parts of the plant aboveground (aerial parts), whereas the phloem cells conduct the sugary products of photosynthesis. The less specialized tissues of the plant are made up mostly of parenchyma cells, where water and food are stored and where photosynthesis takes place.

Leaf structure

In the Venus flytrap the leaf functions both for photosynthesis and as a trap. A waxy cuticle on the leaf surface helps waterproof and protect the leaf. Underneath the cuticle is a single layer of cells called the epidermis, below which is the upper cortex. The center of the leaf has large-celled medullary tissue, with many air spaces between the cells. Below that is the lower cortex, which is made of loosely packed cells, and then the lower epidermis.

Dotted within the lower epidermis are pores called stomata (singular, stoma). Each stoma has two kidney-shape guard cells, which change their shape to open and close the stoma. In this way, they control gas exchange and water loss. Air spaces between the loosely packed lower cortex cells create channels from the stomata, allowing air to diffuse into the leaf tissue.

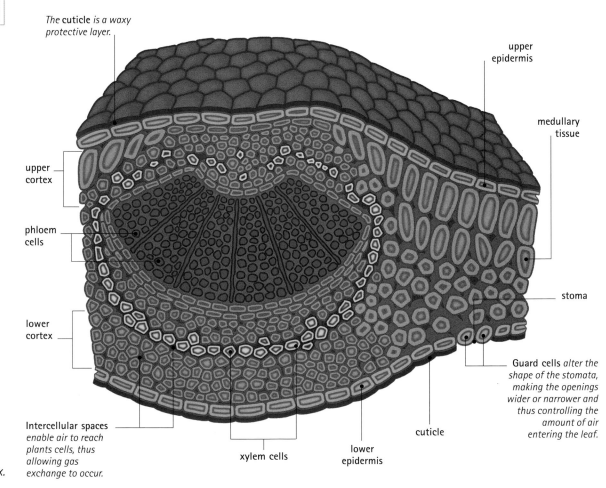

The **cuticle** *is a waxy protective layer.*

upper epidermis

medullary tissue

upper cortex

phloem cells

lower cortex

stoma

Guard cells *alter the shape of the stomata, making the openings wider or narrower and thus controlling the amount of air entering the leaf.*

cuticle

lower epidermis

xylem cells

Intercellular spaces *enable air to reach plants cells, thus allowing gas exchange to occur.*

▶ CROSS SECTION OF LEAF

A waxy cuticle on the outside of the leaf protects and waterproofs it. The outer layer of cells is called the epidermis. This layer surrounds the photosynthesizing tissues: the upper cortex, medullary tissue, and lower cortex.

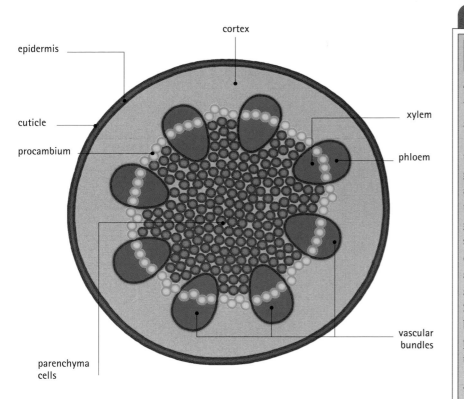

epidermis

cuticle

procambium

cortex

xylem

phloem

vascular
bundles

parenchyma
cells

Photosynthesis, which needs both water and carbon dioxide from the air, takes place in organelles (mini-organs) called chloroplasts that are packed into cells in the medullary tissue. The vascular bundles in the leaves form the veins that supply water and remove the sugary products of photosynthesis. In the Venus flytrap they also transport the nutrients absorbed from digested victims. Between the two leaf lobes is the central midrib. This is the region that folds when the trap is activated. Cells in the midrib are arranged in a zigzag fashion and are capable of rapidly changing shape.

Root structure

In the Venus flytrap's roots—like those of most other dicots—the vascular tissue forms a core through the center. The xylem forms the center of the core with phloem clusters outside this central xylem. Encircling the conductive tissue is a layer of cells called the pericycle. This is the region where side roots are initiated (roots do not produce buds). As they develop, side roots push their way through the cortex and epidermis. Water and minerals from the soil are absorbed mainly from threadlike extensions of epidermal cells called root hairs.

▲ CROSS SECTION
OF RHIZOME
Rhizomes are stems, not roots. Rhizomes have vascular tissue arranged in a ring, around a core of parenchyma, or storage, cells.

Glands

There are two types of glands on the Venus flytrap leaf: those that secrete nectar, and those that secrete the digestive fluid.

The nectar glands are in pits. When the two sides of the leaf are brought together, the glands do not interfere with the airtight seal around the victim. The digestive glands sit on the epidermis, like tiny inverted cones. Each gland consists of about 32 cells with a disk-shape head, a short stalk, and a basal cell that connects it to the rest of the leaf. Although all glands have the same basic structure, they are modified in different carnivorous plants. Sundews have glands on the end of long, bendable stalks. These glands secrete a permanent blob of sticky liquid, waiting to trap unwary insects.

▼ *The round-leaf sundew is native to Europe, North America, and northern Asia. Insects are trapped by sticking to the gluelike substance secreted at the end of the stalks, or tentacles.*

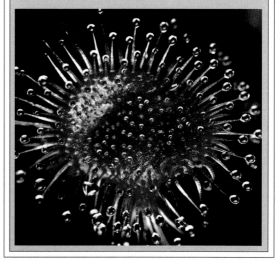

Rhizome structure

The Venus flytrap does not have aerial stems, but it does have underground stems called rhizomes. Their internal anatomy is different from that of the roots. Roots have a central core of vascular tissue, whereas rhizomes—like other stems—have their vascular tissue in a ring, surrounding a core of parenchyma. Branches arise from buds that are either at the tip (terminal branches) or arising from the sides (lateral branches).

Digestive system

The carnivorous plants are all green and so can photosynthesize. Why then do they need to supplement their diet with animals? Most carnivorous plants live in waterlogged areas where the soil is very acidic, and so the bacteria that normally recycle dead plant and animal matter cannot function. In this environment nitrogen and other nutrients are not easily available for plants. Insect bodies are rich in these nutrients, so some plants have evolved over millions of years a way to cut out the recyclers and take the animal nutrients directly.

Despite what is sometimes depicted in horror films, no carnivorous plants are large enough to eat humans. Some of the larger pitcher plants, such as *Nepenthes rajah*, have been known to trap unfortunate rodents and lizards. Most, however, capture insects and other small organisms, down to the size of protozoans and

water fleas. The Venus flytrap does not actually eat many flies. Ants and spiders form about 60 percent of the plant's diet.

The trap snaps shut

The Venus flytrap can shut in about 0.1 second. This closure is one of the fastest movements in the plant kingdom. The leaf margins secrete a sugary substance called nectar, so an insect moves over the leaf surface while it is feeding. The three trigger hairs, which must be knocked to activate the trap, are toward the center. The leaf trap snaps shut only when two hairs are triggered in quick succession (within about 20 seconds), or when one hair is triggered twice. This makes it unlikely that falling detritus will trigger the mechanism, but an insect wandering on the leaf will.

The leaf can shut very rapidly, in 0.1 to 1 second, depending on its size, age, and

▶ **THE CLOSING TRAP**
(1) a robber fly crawls over the open lobes of the Venus flytrap. (2) The fly touches at least two of the trigger hair cells or the same one in succession. (3) Within 1 second, the lobes close, the teeth interlock, and the fly is trapped. The two lobes eventually close to make an airtight seal, and the soft parts of the fly are digested and absorbed.

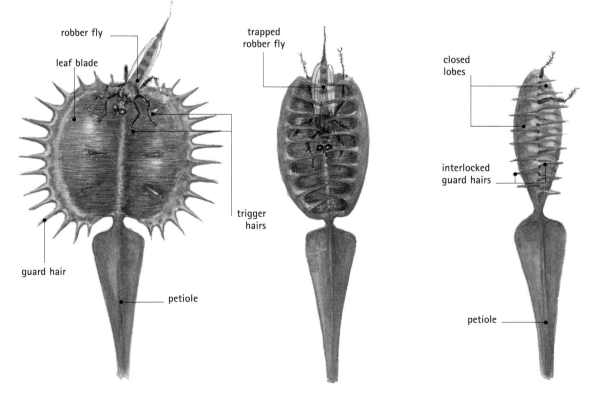

1. A robber fly lands on the leaf blade and touches three of the trigger hairs.

2. The two lobes of the leaf blade shut, trapping the fly. The trap will not close fully for about 30 minues, thus allowing small insects to escape.

3. The robber fly is not small enough to escape, and eventually the lobes close fully. Now the plant can begin to digest its meal.

condition. As the sides of the trap close over their victim, the angled marginal teeth interlock to form a cage. The trap then pauses, almost closed, for a period of about 30 minutes. During this time, if the trap has caught a tiny insect, it can make an escape. This pause prevents the trap from wasting a digestive cycle on a small insect, because each trap can go through only two or three cycles before it is exhausted.

If the insect is still inside and struggling, the second phase of trap closure begins. The rim of the trap presses down to make an airtight seal around the victim. The walls then squeeze with such force that soft-bodied insects can be squashed. This phase is relatively slow. Ants are often still be alive eight hours after a trap has closed on them.

How does the trap work?

The rapid movement of the flytrap has been studied for centuries, but biologists still do not know the exact mechanism of closure. At the large scale, recent research has shown that three-dimensional geometry can explain the

Predators of protozoans

There are several species of plants in the genus Genlisea. These plants, which live in parts of South America and Africa, usually grow in substrates that are very deficient in nutrients, such as moist sand. The plants bear delicate flowers aboveground and bundles of rootlike organs up to 6 inches (15 cm) long underground. The organs are actually leaves, not roots. Botanists had long wondered how *Genlisea* gained the nutrients it needed to stay alive, and the famous naturalist Charles Darwin suggested in 1875 that *Genlisea* was carnivorous. More than a century later, Wilhelm Barthlott, a botanist at the University of Bonn in Germany, confirmed that Darwin was right. Tiny protozoans are attracted—probably by chemicals secreted by the plant—toward slits in the subterranean leaves of *Genlisea*. Once the protozoans, including at least nine species of ciliates, have entered a slit they become trapped and are then digested.

fastest part of the reaction. A physical process called "snap-buckling" is responsible for this rather than a biological, or growth, change. The shape of the two lobes and the tensions across their surface result in their being held in a position that is only on the verge of stability. In the open state, the lobes have a convex (outward-curved) shape. A minor change in the shape of the midrib, situated between the lobes, leads to extra tension that pushes the leaf over its "stability threshold." The lobes then snap into their more stable position in which they have a concave (inwardly curved) shape and are bent together.

The trigger for snap-buckling is thought to be rapid growth (stretching) of cells along the outer edge of the midrib. One theory is that

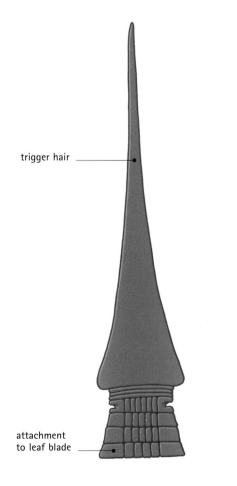

trigger hair

attachment to leaf blade

◀ TRIGGER HAIR
When two trigger hairs are touched within about 20 seconds, or when one is triggered twice, the leaf trap snaps shut.

▲ *Ants and spiders, rather than flies, make up the majority of the Venus flytrap's diet.*

the stimulus triggers a change in the membrane of cells, possibly acid release, softening the cell wall. The pressure of water inside the cells (turgor pressure) causes them to swell rapidly as soon as the wall is stretchy enough. Because the outer edges of the midrib grow more than those on the inside, the leaf curves just enough to trigger the fast snap-buckling.

Cell signaling

At a cellular level, the changes that create this snap-buckling occur by electrical signals called action potentials. Every cell maintains a slight difference between the ions (for example, potassium and chloride ions) that are inside and outside the cell. Certain ions are actively pumped into or out of the cell. Even a tiny

difference between the numbers of positive and negative ions on either side of the cell membrane creates an electrical gradient across the cell membrane. An action potential occurs when something destabilizes the cell membrane, and the gradient is temporarily lost. The change can then be passed on to an adjacent cell, resulting in an electrical signal passing through a tissue. Animals have a similar electrical communication system, but their electrical impulses travel along nerve cells, or neurons. These cells form networks, linked at junctions called synapses.

Plants, however, do not have a specialized nervous system. Instead, the electrical impulses pass between ordinary cells. Plant cells are boxed in by their thick cell walls, but there are tiny gaps called plasmodesmata between adjoining cells. Cytoplasm (the cell contents) connects the cells through these gaps, so this is the route that electrical signals in plant cells have to take.

Insect soup

Once the trap has sealed, the digestive process begins. Like salivary glands in a mammal, digestive glands in the Venus flytrap secrete fluid only when stimulated. In a mammal, the stimulation might be the scent of food. In the Venus flytrap, it is the sensation of a trapped insect struggling, together with chemicals that it releases. Urea, an insect excretory product, is a powerful stimulant, as are amino acids and sodium ions—these are both stimulants that are present at concentrations similar to those found in insect hemolymph (bloodlike fluid in an insect).

The digestive fluid comes from the reddish secretory glands. It contains protease enzymes (complex chemicals that digest proteins). The fluid also contains some hydrochloric acid. The digestive juices of pitcher plants are extremely acidic, at pH 2 to 3. This acidity is

similar to that in the human stomach. The enzymes in the Venus flytrap's digestive fluid turn the soft parts of the victim to liquid. This insect soup is then absorbed by the plant. The whole process, from the closure of the trap to absorption, takes about two weeks. The lobes of the trap then slowly open to reveal the insect's wings and husk (the indigestible exoskeleton), which blow or wash away.

COMPARATIVE ANATOMY

Bacterial helpers

Pitcher plants use bacteria to help them digest their prey. The bacteria secrete enzymes that break down the prey components into a soup. The bacteria use the breakdown products, but plenty is left for the plant, too. Mammals also have bacteria in their digestive system. Herbivores that eat plants have a problem. Cellulose—the main component of plant cell walls—is very tough, and only the enzyme cellulase can break it down. Mammals cannot make cellulase. Instead, bacteria that can produce the enzyme live in the digestive tract and digest the plant matter. The animal then digests the breakdown products, and some of the bacteria, too.

▼ *Once a sizable victim is trapped, the Venus flytrap takes about two weeks to break down the animal's soft parts and absorb the nitrogen-rich "soup."*

Reproductive system

CONNECTIONS

COMPARE the Venus flytrap with *MARSH GRASS*. Both can spread over relatively large areas of soil using underground stems called rhizomes. However, to reproduce sexually they use different pollination methods. Marsh grass is pollinated by the wind, whereas the Venus flytrap is pollinated by insects.

In its natural habitat, the Venus flytrap spreads vegetatively, by producing new plants that bud from the underground stems, or rhizomes. In this form of asexual reproduction, the plants can cover large areas of suitable soil. However, all plants that are produced by asexual reproduction are clones of their parents: they are genetically identical. Sexual reproduction is by way of flowers and seeds. The separation of chromosomes (which carry genes), occasional mutations of genes, and the random pairing of male pollen grains with female ovules ensure that combinations of genes are "shuffled." This mixing produces genetic variation, which ensures that at least a few offspring stand a chance of surviving when environmental conditions that may once have suited the parent plant change.

IN FOCUS

Don't eat the pollinators!

When a Venus flytrap flowers, it produces a tall flower stem of up to 18 inches (45 cm) high from the center of a rosette, which is only about 6 inches (15 cm) wide. An extra-long flower stalk is common in many carnivorous plants. It holds the flowers safely away from the dangerous traps at the base, because the plant would not benefit from trapping an insect that is carrying its pollen before the insect has had the chance to transfer pollen to another flower.

◄ *A white-legged damselfly has become trapped by the great, or English, sundew. As the insect struggles, more of the plant's sticky tentacles curl onto it.*

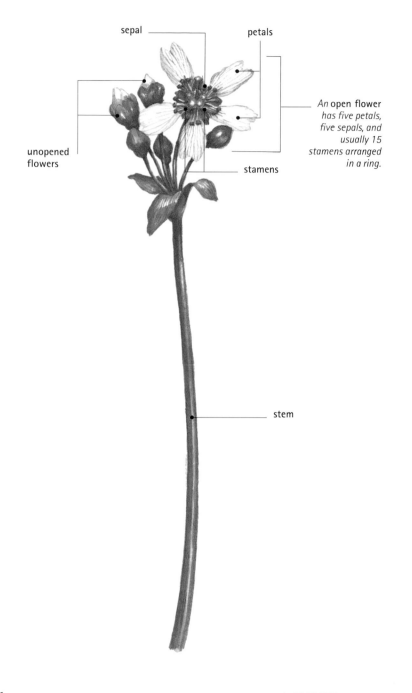

sepal

petals

unopened flowers

An open flower has five petals, five sepals, and usually 15 stamens arranged in a ring.

stamens

stem

▲ FLOWER
Venus flytrap
Clusters of small white flowers reach 18 inches (45 cm) high. The flowers are pollinated by bees and other insects.

Flower structure

In the wild, the Venus flytrap flowers in May and June. A tall flower stem grows from the center of the rosette, bearing up to 15 buds. The buds open to reveal white flowers. The flower has radial symmetry (if the flower is rotated around its center, there is more than one position during a single rotation in which the flower looks the same as at the starting point). Four elements make up a flower: the sepals (the green bracts that protected the unopened flower bud); the petals; the stamens (male parts); and the pistil (female parts).

Pollination and fertilization

Pollen is carried from one flower to another by bees and other insects. When pollen lands on a stigma, the pollen germinates and produces a pollen tube. The tube penetrates the stigma and grows down the style, into the ovary. When the tube eventually pierces an ovule, the nuclei of the pollen and ovule unite. The fertilized ovule then forms the embryo of the seed.

By late July, the ovary wall has swollen and dried, becoming a seed capsule filled with lots of small, slippery, pear-shape black seeds. The capsule splits, scattering the seeds around the parent plant. They germinate soon after release, so the young plant has a few months to grow before its first winter. It takes three or four years for the new Venus flytrap plant to flower.

The arrangement of the floral elements is relatively simple with five sepals, five white petals, numerous stamens (usually 15), and one pistil. The elements occur in rings (whorls) with the sepals on the outside and the pistil on the inside, all growing from the receptacle (the tissue that forms the base of the flower).

The stamen has a long filament, on top of which is the anther, which contains the pollen. The pistil consists of the swollen ovary containing a single chamber (locule) that houses many ovules, possibly 100 or more. The ovules, if fertilized, form the seeds. Above the ovary is the stalklike style, which ends in a receptive surface called the stigma, on which pollen lands and germinates.

ERICA BOWER

FURTHER READING AND RESEARCH

D'Amato, P. 1998. *The Savage Garden: Cultivating Carnivorous Plants.* Ten Speed Press: Berkeley, CA.
Pietropaolo, J., and P. Pietropaolo. 1996. *Carnivorous Plants of the World.* Timber Press: Portland, OR.
Carnivorous Plants:
    waynesword.palomar.edu/carnivor.htm
International Carnivorous Plant Society:
    www.carnivorousplants.org

Virus

Viruses are tiny but extremely successful parasites that multiply inside the cells of other living organisms. They are so small that their anatomy is in terms of only atoms and molecules, not cells or organs. In fact, a simple virus consists of little more than a set of genetic instructions in the form of the nucleic acids DNA or RNA, wrapped up in a protective layer of protein.

The cells of all living things—from bacteria to humans—are at risk of attack from viruses. In humans, viruses cause serious diseases such as influenza, smallpox, and AIDS (acquired immune deficiency syndrome). Complete virus particles, called virions, normally exist only when the virus is outside a cell. Virions cannot move themselves around and are so inert (lacking in any metabolic activity) that viruses are sometimes claimed not to be truly living. They rely on outside forces to reach new hosts. For example, common cold viruses are inhaled in water droplets produced by coughing or sneezing. Others are spread by food or drink or by the act of sexual reproduction. Plant viruses are often spread by insects.

When a virus reaches a cell there are various ways in which the virus can invade it. Insects often inject plant viruses directly into cells, as a side effect of feeding on the plants. Many viruses attack only particular kinds of cells and have molecules on their surfaces that help them recognize the right type of cell to invade. Sometimes, it is just the genes of the virus that cross into the cell. These genes then begin the process of hijacking the cell's own anatomical structures to make more viruses.

Classifying viruses

Viruses come in various shapes and sizes and are classified into at least 70 different families. Examples include the poxvirus family (the smallpox virus and its relatives) and the retrovirus family (which includes the virus that causes AIDS). Within a family, all the viruses are probably related to one another. However, no overall family tree can be drawn for all viruses. This is probably because different viruses originated in different ways. For example, it is possible that the largest viruses, such as the poxviruses, were originally cells that degenerated, whereas this origin is unlikely for smaller viruses.

For practical purposes, scientists classify viruses in various ways. The simplest way is to divide them into groups according to the type of organisms that they infect: animal, plant, and bacterial viruses. However, not all animal viruses, for example, are closely related to one another. Viruses are also classified by their size and shape.

A particularly important way of classifying viruses is by the type of genetic material (DNA or RNA) they have. Viruses show much more variety in this respect than cells,

▼ Scientists classify viruses in various ways: for example, according to their type of genetic material, by the disorders that they cause, or by their size or shape. The table below shows various types of viruses that cause disorders in humans.

DNA viruses that cause human disorders

FAMILY	GENUS	DISORDER
Herpesviridae	Simplexvirus	Cold sores, encephalitis
Poxviridae	Orthopoxvirus	Smallpox
Adenoviridae	Mastadenovirus	Some pneumonia, conjunctivitis, cystitis
Papovaviridae	Papillomavirus	Warts, cervical carcinoma

RNA viruses that cause human disorders

FAMILY	GENUS	DISORDER
Picornaviridae	Enterovirus and others	Poliomyelitis, most common cold
Togaviridae	Rubivirus	Rubella
Coronaviridae	Coronavirus	Some common cold
Rhabdoviridae	Lyssavirus	Rabies
Arenaviridae	Arenavirus	Lassa fever
Orthomyxoviridae	Influenzavirus	Influenza
Retroviridae	Lentivirus	AIDS

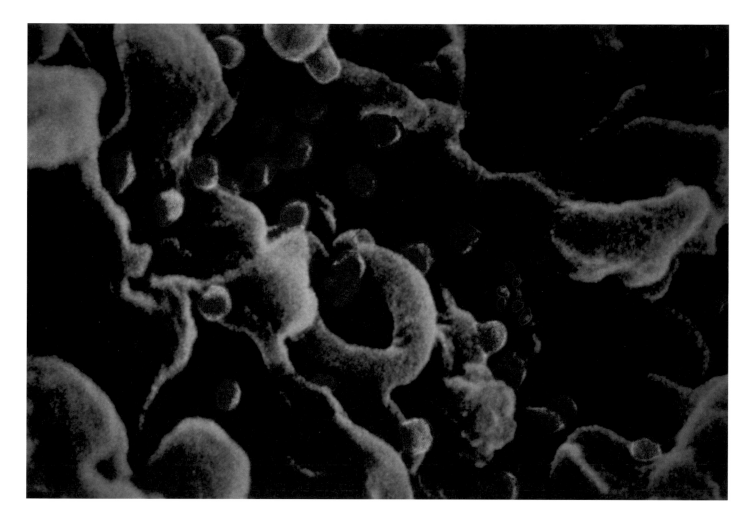

which all carry their genes in the form of a double strand of DNA (the famous double helix). The genes of different viruses can be in the form of single- or double-stranded DNA or RNA, depending on the virus. These distinctions form the basis of the Baltimore classification of viruses, named for U.S. virologist David Baltimore. It is a useful classification because the nature of a virus's genes affects

▲ The red particles in this false-color transmission electron micrograph (TEM) are human immunodeficiency viruses (HIV, the cause of AIDS), which are shown here attacking white blood cells called CD4 lymphocytes. (x 85,000 magnification.)

the way in which it behaves inside a cell. For example, viruses with genes in the form of DNA tend to multiply in the host cell's nucleus because it is there that the cell's own DNA-copying machinery is found.

Impact of viruses

Because viruses can cause serious diseases, they are of huge concern to humanity. In addition to diseases in humans, they cause diseases of livestock and agricultural crops. A vast amount of scientific research centers on developing drugs that attack viruses or vaccines that prevent them from invading the body. However, viruses can evolve quickly, and it seems unlikely that we will ever get rid of viral diseases completely. Viruses have also become useful tools for the scientific researcher. Much of our modern understanding of modern genetics and cell biology comes from studying viruses. In addition, viruses have been vital in the development of genetic engineering techniques.

FEATURED SYSTEMS

STRUCTURE OF VIRUSES Viruses are neatly constructed for their role of invading cells. The main components of viruses are genetic material (DNA or RNA), a protein coat (capsid), and in some an outer envelope made of lipids. *See pages 1316–1318.*

VIRUSES WITHIN CELLS Viruses have various ways in which they take over cells for their own purposes. *See pages 1319–1321.*

DNA VIRUSES Some of the largest viruses use DNA, not RNA, as their genetic material. *See page 1322.*

RETROVIRUSES The anatomy of these unique viruses has become much better known since the AIDS epidemic began. *See pages 1323.*

Structure of viruses

COMPARE the simple structure of viruses with that of *BACTERIA*. A virus's principal components are DNA or RNA, a protein coat (capsid), and in some a lipid envelope. In contrast, bacteria are larger and contain a more complex internal structure.

The structure of a complete viral particle (virion) varies greatly between different families of viruses. All viruses, though, have at least one molecule of either DNA or RNA at their center. The DNA or RNA contains the virus's genes. In the simplest viruses, the genes are surrounded by only a single protective coating called a capsid, which is made up of many identical protein molecules packed together. Each protein molecule is long and thin but curls up into a ball-like shape. The curled-up molecules fit together side by side, like bricks in a wall.

All viruses are much smaller than the cells they parasitize. A typical animal cell might be 20 microns (1/50 millimeter) long, but some small viruses are 1,000 times smaller than this—only 20 nanometers in diameter.

Discovering viral structure

How do we know so much about such tiny objects as viruses? Before the 1930s, no one had even seen a virus. Then, the invention of the electron microscope allowed scientists to see the shapes of viruses for the first time. The geometrical shapes of many viruses allow them to be made—by a chemical process—into a crystalline form, like salt, so that their shape can be figured out using X rays. Many chemical methods have also been used to analyze viruses and to read the messages stored in their genes.

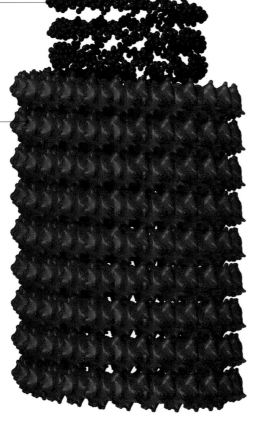

single strand of RNA

The cylindrical **capsid** *is made up of many copies of a single repeating protein.*

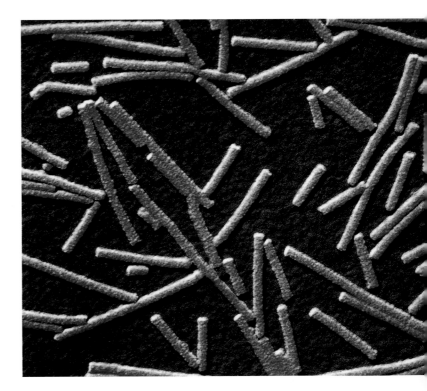

◀ Tobacco mosaic virus (TMV) *The structure of TMV is particularly simple: a cylindrical capsid of more than 2,000 copies of the same protein surrounding a molecule of RNA.*

▲ *The rod-shape tobacco mosaic virus was the first virus to be discovered. It infects many plants, causing distortion and mottling of the leaves. (x 250,000 magnification.)*

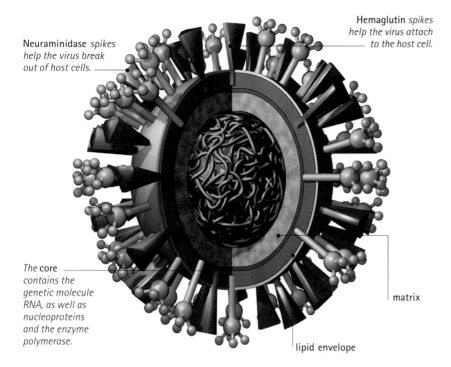

Neuraminidase *spikes help the virus break out of host cells.*

Hemaglutin *spikes help the virus attach to the host cell.*

◀ **Influenza virion**

The human influenza virus has a ball-like shape and contains RNA. The virion has an envelope with molecules that help the virus infect cells and molecules that help new viruses break out of cells.

The **core** *contains the genetic molecule RNA, as well as nucleoproteins and the enzyme polymerase.*

matrix

lipid envelope

Viruses that have only a simple capsid have geometrical shapes. In some families, the viruses also have an extra envelope made mainly of lipids (fatty molecules). Such enveloped viruses are often spherical, but they come in a variety of shapes. For example, enveloped viruses of the rabies virus family have bulletlike shapes. The largest viruses, such as the poxviruses, have a more complex anatomy.

Outer coverings

A virus's outer coat protects it, contains the molecules inside, and often helps it recognize and invade living cells. The first virus to have its outer covering well understood was the

▼ **Icosahedral capsid**

Many animal viruses have a 20-faced, or icosahedral, capsid. The capsid is made up of ring-shaped units called capsomers. There are two types of capsomers: pentons, which have five subunits, and hexons, which have six.

The core of the capsid is called the **nucleocapsid.**

tobacco mosaic virus (TMV). This much-studied virus of plants has a fairly simple structure, consisting of a cylindrical capsid that surrounds a single molecule of RNA. The capsid consists of more than 2,000 copies of the same protein, packed tightly together in a spiral arrangement to form a cylinder. Because only one kind of protein is used, the virus needs only one gene for its capsid—a single gene typically contains the blueprint for making just one kind of protein.

The shape of TMV's protein molecules and the attraction between them result in their natural self-assembly into cylinders when they are mixed with the correct type of RNA, even in a test tube. (In nature, this self-assembly occurs inside the tobacco plant cells that the virus attacks.)

Many viruses have a cylindrical capsid, but an even more common shape—especially with animal viruses—is an icosahedron. This is a symmetrical ball-like shape with 20 flat sides. Like a cylindrical capsid, an icosahedron has

Capsomers with six subunits are called **hexons.**

IN FOCUS

Viroids

There are parasites called viroids that are even smaller than viruses. They do not have a protein coat and consist of only a single short molecule of RNA curled up into a three-dimensional shape. Viroids contain no genetic instructions for making anything; they are just very good at getting cells to make more copies of themselves. Most viroids infect plants, and some cause serious plant diseases.

Capsomers with five subunits are called **pentons.** *This capsid has 12 pentons.*

This virus has 42 **capsomers,** *but other icosahedral viruses have many more.*

Dangerous proteins

Scientists were amazed to discover infectious particles that appear just to be proteins and contain no genes. Called prions, these are the likely cause of bovine spongiform encephalopathy (commonly called mad cow disease) in cattle and a similar brain disease called Creutzfeldt-Jakob disease suffered by humans who have eaten infected beef. Prions appear to be natural cell proteins that are folded up in a different way from their normal form. Somehow, if a prion protein enters a nerve cell, it seems able to cause ordinary proteins to change shape into the prion shape, causing disease as a result.

the advantage that it can self-assemble from smaller protein units (either one kind or several kinds). In viruses in which the whole capsid is designed to invade a cell, the capsid also needs the property of being able to disassemble at the right moment, to release the virus's genes so they can begin their work.

Some viruses also have a flexible membrane called an envelope outside the protein capsid. The envelope is made of lipids that are usually stolen from a membrane of the cell in which the virus was originally assembled. Dotted among the lipids of the envelope of these viruses are other molecules specific to that virus. Particularly important are glycoproteins, which are molecules that are a combination of sugars and proteins. Glycoproteins often stick out as spikes from the envelope. These molecules help viruses recognize cells that are suitable for invasion and trick the host cell into letting the virus infect the cell.

Inside story

Many virions contain extra proteins, in addition to those in the capsids and envelopes. These proteins are often enzymes—biological catalysts that promote particular chemical reactions, such as copying DNA or RNA.

One thing that is always found inside a virus is its genome—that is, its genetic instructions, in the form of a nucleic acid (either DNA or RNA). The genome is tightly packed within the virion, often wound around proteins. In some viruses, there is a single molecule of nucleic acid. In others, there are several molecules of nucleic acid. In some plant viruses the whole genome is too big to fit into a single particle. In such viruses, a successful infection needs more than one virus particle, each containing a different part of the genome.

The number of genes in a virus varies from three or four to 200 or more, depending on the family of viruses. The larger the virus, the more genes it tends to have. Some viruses have elaborate methods for taking over the cell's machinery, and these require extra genetic instructions. For the viruses that attack large organisms such as humans, some genes may act to neutralize the body's defense systems for repelling diseases.

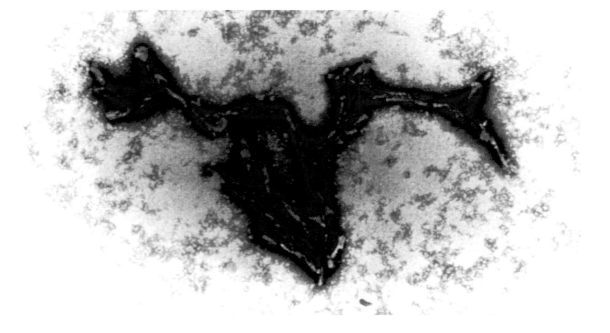

▶ *This false-color transmission electron micrograph (TEM) shows prion proteins. Prions are infectious particles with no genetic material; they can change normal proteins in the brain into an abnormal form. Prions damage nerves, leading to brain degeneration, and are a cause of the human brain disorder Creutzfeldt-Jakob disease (CJD).*

Viruses within cells

Inside cells, viruses spring into action. Like other parasitic life-forms, they have special features that allow them to survive and reproduce within other living organisms. In the case of viruses, a host cell's anatomy and biochemical systems are crucial to the virus's success—in effect, the virus hijacks a cell to serve its own purposes.

Different types of viruses have different ways of operating inside a cell. The most common way is for a virus to take over a cell completely. A single invading virus can cause a cell to make thousands of copies of the same virus, before the cell eventually dies and the new viruses are released. Some other viruses do not do this immediately. Instead, their genes remain shut off and divide only when the cell divides, although eventually they become activated and produce more viruses. Yet other viruses constantly shed copies of themselves from an infected cell but without killing the cell itself.

Gene processing

Once the genes of a virus enter a cell, they have two main tasks. They must arrange for the various instructions they contain to be put into action, and they must also arrange for copies of themselves to be made for packaging into new viruses. To understand how viral genes perform these tasks, it is necessary to understand how a cell's own genes produce their effects and also how they copy themselves.

In animal and plant cells, the genes are made of DNA and are housed in the cell nucleus. An individual gene typically contains the blueprint for making a particular kind of protein. In some cases, these proteins form cell structures or enzymes; in other cases, their job is to

▼ *Avian flu is caused by a type of influenza virus. The ball-like virus infects all birds, and may spread to humans. A disease that can spread from an animal to a human is called a zoonosis.*

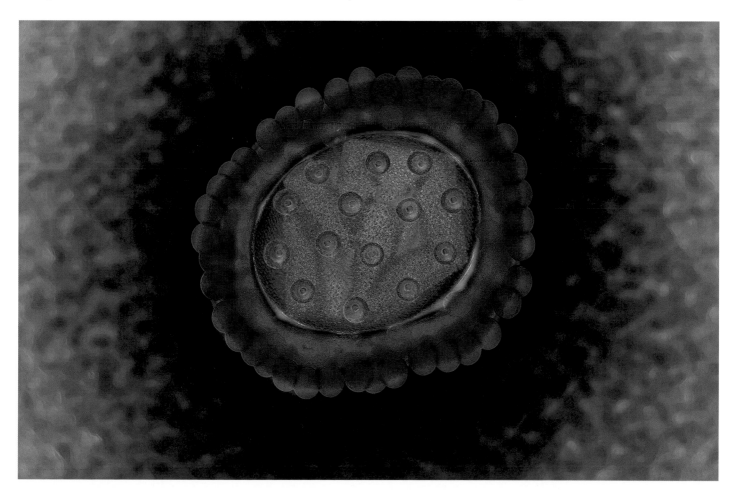

► *This false-color scanning electron micrograph (SEM) shows T-bacteriophage viruses attacking a cell of the bacterium* Escherichia coli, *which is commonly found in the intestine of animals. The virus attaches its tail to the cell wall of the bacterium and "squirts" its DNA into the host cell. (x 50,000 magnification.)*

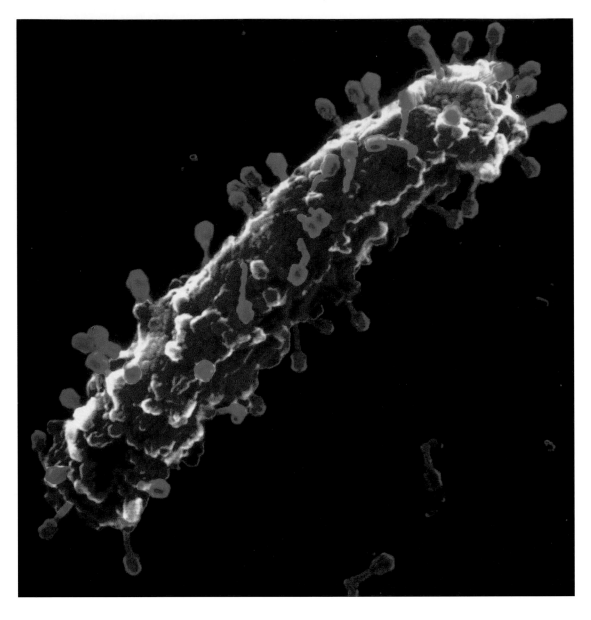

switch other genes on or off in a complex system of control. *Gene expression* is the term for the way that a gene produces its effects.

For a cell's own genes, gene expression is a two-stage process. First, an enzyme in the nucleus copies a particular gene from DNA into RNA—a process called transcription. This RNA, called messenger RNA, then leaves the nucleus and enters the cytoplasm—the gel-like material surrounding the nucleus. In the cytoplasm, large molecular structures called ribosomes "read" the code in the RNA and make a particular protein that matches the code. This process is called translation.

When a virus enters a cell, it takes over and makes use of this cellular system. The exact details depend on the type of virus—especially whether it is a DNA or RNA virus. RNA viruses usually trick cells into treating their genes as if they were the cell's messenger RNA, and so the cell then translates the viral RNA into viral proteins.

Viruses usually have a number of different genes, and the proteins they produce have different effects on their host cell. Many viruses have a gene that makes a protein that shuts down the host cell's own protein synthesis (manufacture), so only viral proteins are made. The virus can also influence the cell so that different viral proteins are made at different times. For example, viral-coat proteins tend to be made at a later stage of cell infection, when it is time for the viruses to assemble and break out of the doomed cell.

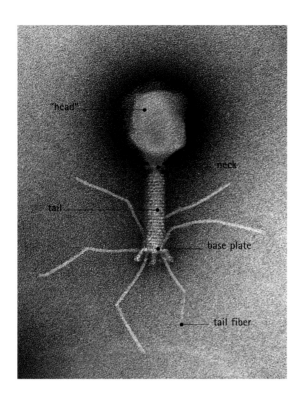

▲ *A T-bacteriophage virus has an icosahedral head, which contains genetic material; and a contractile tail, which attaches itself to the host bacterium. (x 110,000 magnification.)*

For making copies of their own genes to put into new viruses, RNA viruses have a basic problem. Cells do not normally contain an enzyme that will copy RNA into more RNA. So RNA viruses do one of two things. Either they bring in existing enzymes with them when they invade the cell, or they have a gene for an RNA-copying enzyme, which they "persuade" the cell to produce for them. DNA viruses mostly use the cell's own DNA-copying machinery. The genes of DNA viruses therefore tend to make their way to the cell nucleus and operate from there.

Assembly lines

Assembling a complete virus is a complex operation, and details vary. Although capsids can self-assemble, this ability does not solve the problem of getting things into the right place at the right time. It turns out that the cell has an addressing system used to direct its own molecules to the right places, and viruses take advantage of this system.

For example, in herpesviruses, which are DNA viruses, the viral genes are transcribed into messenger RNA in the nucleus by the cell's enzymes. The messenger RNA then passes into the cytoplasm, where, for example, the capsid proteins are made. However, the capsid proteins then have to get back into the nucleus to assemble around the genes. In other words, the virus has to "navigate" its way around the cell's anatomy. In this case, the capsid proteins contain a pattern that the cell interprets as: "Transport me to the nucleus!" This signal—which is used by all cells—was first discovered by studying viruses.

The later stage of assembly also has various patterns. In some cases, the cell's secretory systems are hijacked to transport the finished viruses to the outside of the cell. Cell membranes are also taken by viruses for use as envelopes. Final assembly often takes place near the cell's outer membrane. Further changes, such as the chopping up of proteins into their final shapes, may occur even after the virus particle is released. Assembling a virus is not necessarily efficient. With some viruses, many incomplete virus particles are left behind in the host cell. However, enough complete viruses are usually produced for a viral infection cycle to begin anew.

▼ Picornavirus life cycle

Picornaviruses include polioviruses and many common cold viruses. The main stages of a typical picornaviral life cycle are shown below. Picornaviruses are RNA viruses and do not invade the cell nucleus.

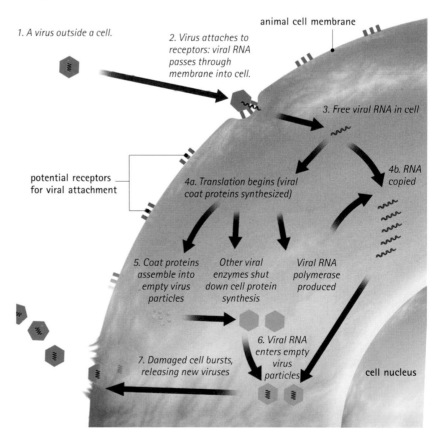

1. A virus outside a cell.

2. Virus attaches to receptors: viral RNA passes through membrane into cell.

animal cell membrane

3. Free viral RNA in cell

4b. RNA copied

potential receptors for viral attachment

4a. Translation begins (viral coat proteins synthesized)

5. Coat proteins assemble into empty virus particles

Other viral enzymes shut down cell protein synthesis

Viral RNA polymerase produced

6. Viral RNA enters empty virus particles

7. Damaged cell bursts, releasing new viruses

cell nucleus

DNA viruses

Most viruses have genes made of RNA, but some of the largest and most interesting viruses use DNA instead, in the same way that animal and plant cells do. RNA and DNA are structually very similar. However, the structure of DNA lends itself more to forming a double strand, whereas RNA usually occurs as just a single strand (although there are some viruses with double-stranded RNA).

One family of double-stranded DNA viruses is the herpesviruses, which cause various diseases, including: chicken pox in children and its "adult" form, shingles; infectious mononucleosis; and herpes infections themselves (cold sores and genital herpes). These are typical DNA viruses, in that they operate and reproduce themselves mainly in a cell nucleus. An unusual feature of herpesviruses is that they can become latent (inactive) in the body's neurons (nerve cells), only to break out into an infection later. For this reason, and because there are no effective vaccines against them, herpesviruses are very difficult to control.

▼ **Herpesvirus**
Herpesviruses cause disorders such as chicken pox, cold sores, and genital herpes. These viruses operate and make copies of themselves mainly in the cell nucleus.

Bacteria eaters

Viruses that attack bacteria are called bacteriophages, which means "bacteria eaters." Both DNA and RNA viruses attack bacteria, but the best-studied bacteriophages are large DNA viruses whose shape is a combination of an icosahedral head and a cylindrical tail. Many important discoveries about viruses and the nature of genes in general were first made by scientists studying bacteriophages.

▼ *Bacteriophages have mostly double-stranded DNA similar to all bacterial chromosomes (below), which they inject into their host, the bacterium E. coli. (x 56,500 magnification.)*

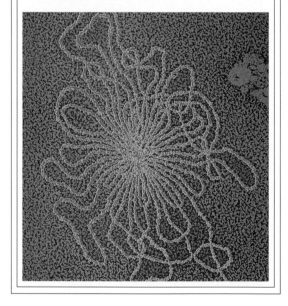

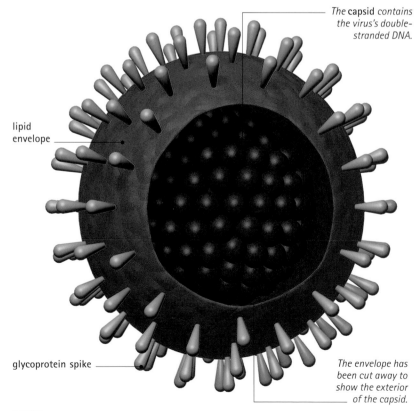

*The **capsid** contains the virus's double-stranded DNA.*

lipid envelope

glycoprotein spike

The envelope has been cut away to show the exterior of the capsid.

The largest of all viruses are the poxviruses, which include the virus that causes smallpox. They are so big (in viral terms) that they can just be seen under an ordinary light microscope. Their genome contains about 200 genes. The poxviruses carry all their own DNA-copying machinery, and so do not need to operate from the infected cell's nucleus. Poxviruses have a complex structure, which scientists do not yet fully understand; it includes many different kinds of proteins.

Retroviruses

Retroviruses make up a unique family of viruses that includes human immunodeficiency virus (HIV), the cause of the disorder AIDS. Most known members of this virus family infect vertebrates (animals with a backbone). All retroviruses have the same basic structure, and all are probably related. They carry their genes as RNA, but unlike all other RNA viruses they "arrange" for their RNA to be copied back into DNA. This DNA then merges with the genes of the cell that they are attacking.

Clever operators

A retrovirus recognizes its target cell by using proteins on the virus's outside surface. These lock onto other proteins that are found on only the types of cells that the virus is designed to invade. In HIV, for example, the host cells have a receptor called CD4 and include a type of white blood cell called a CD4 lymphocyte, which fights infection and is part of the human immune system. The membranes of the cell and virus then fuse, releasing the inner structure of the virus into the cell's cytoplasm.

An enzyme produced by the retrovirus called reverse transcriptase then copies the virus's genes from RNA into DNA. This DNA copy, which is called a provirus, moves to the cell's nucleus, where another viral enzyme inserts the provirus among the host cell's own genes. Once there, the provirus is never removed—it stays there as long as that particular cell survives. Sooner or later, the provirus becomes active. It uses the cell's mechanisms to make RNA, which is used both to make viral proteins and to become the genetic material of new virus particles.

Retroviruses and disease

Some retroviruses (although not usually HIV) can cause cancer. This effect is connected with their merging with the cell's own genes during infection. Sometimes, they may insert into the middle of a control gene of a cell, so the cell starts dividing too often and uncontrollably—that is, it becomes cancerous. Retroviruses can

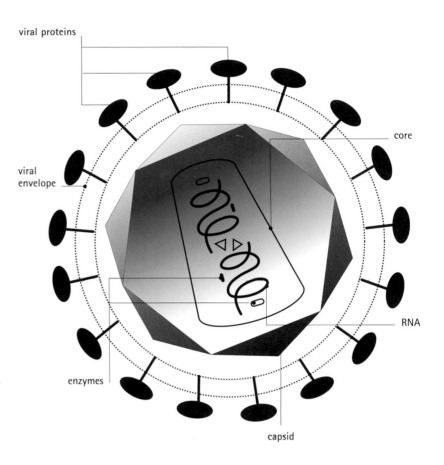

▲ **Human immunodeficiency virus (HIV)**
Like other retroviruses, HIV has an enveloped virion (virus particle). It is surrounded by a membrane (viral envelope) taken from a previous host cell, albeit studded with viral proteins. Within the core there are, uniquely, two complete copies of the viral genome.

also pick up cancer-causing genes from one individual and transfer them to another. The problem with HIV—apart from there being as yet no vaccine against it—is that the virus attacks the body's own defenses, so the body can no longer fight off other diseases, such as opportunistic infections and cancers.

RICHARD BEATTY

FURTHER READING AND RESEARCH
Flint, S. J., V. R. Racaniello, L. W. Enquist, and A. M. Skalka. 2003. *Principles of Virology: Molecular Biology, Pathogenesis, and Control of Animal Viruses*. American Society of Microbiology: Washington, DC.
Preston, R. 1995. *The Hot Zone*. Anchor Books: New York.
Sompayrac, L. 2002. *How Pathogenic Viruses Work*. Jones and Bartlett: Boston.

Weasel

ORDER: Carnivora FAMILY: Mustelidae
SUBFAMILY: Mustelinae GENUS: *Mustela*

Weasels are dedicated hunters with shearing cheek teeth called carnassials; and they include some of the most efficient of all mammalian predators. Weasels are adapted for hunting in burrows, runs, and crevices, with a very long, sinuous body that allows them to slip through tunnels in pursuit of prey. The smallest of the weasel family, the least weasel, occurs throughout most of Europe, northern Asia, and northern North America, including the Arctic. It is suited to hunting small mammals in their burrows and spends much of its time underground or beneath snow. Similar species are found almost worldwide, although those in Australia and New Zealand are descended from animals introduced by humans. There are 16 species of true weasels in the genus *Mustela*.

Anatomy and taxonomy

Animals and other organisms are classified in groups based mainly on shared anatomical features. These usually indicate that the members of a group have the same ancestry, so their taxonomy shows how they are related to one another. However, in recent years studies comparing the DNA of species have revealed relationships that are not reflected in general anatomy, and many taxonomic groupings have been revised as a result.

● **Animals** All true animals are multicellular organisms with well-developed powers of movement, using muscles, and the ability to respond rapidly to environmental changes and events. They obtain nutrients by eating other living things, digesting the tissues into simpler substances. Animals use these substances to build their own body and as a source of energy.

● **Chordates** A chordate has a body that is supported by a pliable reinforcing rod called a notochord. This provides some rigidity and helps the muscles work more effectively.

▼ *The least weasel is one of 16 species of weasels, polecats, ferrets, and minks in the genus* Mustela, *one of nine genera in the weasel subfamily Mustelinae. The subfamily also includes the martens and four species each classified in genera of their own. The subfamily is one of five in the family Mustelidae, which also includes the otters and badgers. The classification within the family is still debated.*

Animals
KINGDOM Animalia

Chordates
PHYLUM Chordata

Vertebrates
SUBPHYLUM Vertebrata

Mammals
CLASS Mammalia

Carnivores
ORDER Carnivora

Catlike carnivores
SUPERFAMILY Feloidea

Dogs, bears, weasels, and raccoons
SUPERFAMILY Canoidea

Mongooses, civets, and hyenas
SUBFAMILY Feliformia

Weasels and relatives
FAMILY Mustelidae

Otters
SUBFAMILY Lutrinae

Weasels, polecats, and martens
SUBFAMILY Mustelinae

Badgers
SUBFAMILY Melinae

Grisons
GENUS *Galictis*

Wolverines
GENUS *Gulo*

Weasels, polecats, ferrets, and minks
GENUS *Mustela*

Tayras
GENUS *Eira*

Martens
GENUS *Martes*

- **Vertebrates** A vertebrate has a notochord that forms the basis of a flexible backbone made up of units called vertebrae. The backbone is the foundation of a skeleton that is made of cartilage or bone. The skeletal units provide anchorage for muscles that are the same on the left and right of the body; this mirror-image arrangement is called bilateral symmetry. A vertebrate also has a skull with a strong cranium enclosing the brain, and the group that the vertebrates form is sometimes called subphylum Craniata for that reason.

▲ *Weasels often hunt animals that are larger than themselves. This least weasel has just killed a rabbit several times its weight.*

- **Mammals** These are vertebrates that feed their young on milk produced by the females. Mammals are warm-blooded animals that have fur or hair at some stage. A mammal's jaw is hinged directly to its skull, unlike the jaws of all other vertebrates, and its red blood cells do not have a nucleus.

- **Placental mammals** Unlike marsupials or monotremes, placental mammals nourish their unborn young during pregnancy with nutrients from the mother's bloodstream. Nutrients from the mother's blood pass across a temporary organ called the placenta, which is attached to the wall of the uterus. From the placenta, the nutrients travel to the young along blood vessels in an umbilical cord.

- **Carnivores** Mammals of the order Carnivora, such as the cats, dogs, and weasels, are mainly meat eaters. Typical carnivores are equipped with carnassials, chewing teeth modified into shearing blades for slicing through flesh.

- **Mustelids** The family Mustelidae is the most diverse group in the order Carnivora. It includes heavyweight diggers and hunters like the American badger and wolverine, the aquatic otters, the tree-climbing martens, and the burrow-hunting weasels. Mustelids tend to have a long body and relatively short legs, and most species are active predators. Exceptions include the European badger, which feeds mainly on earthworms.

- **Weasels, polecats, and martens** The subfamily Mustelinae includes the weasels, polecats, minks, and martens, and the wolverine. There are 33 species.

FEATURED SYSTEMS

EXTERNAL ANATOMY Weasels are slender, short-legged carnivores adapted for hunting underground in burrows. *See pages 1326–1331.*

SKELETAL SYSTEM The bones of the body are slender, since they do not need to bear much weight, but the skull and jaw are very strong. *See pages 1332–1333.*

MUSCULAR SYSTEM Owing to a weasel's small size its muscles exert a lot of power, enabling it to carry heavy loads. *See page 1334.*

NERVOUS SYSTEM Acute senses and a well-developed intelligence allow the weasel to track prey and devise tactics for catching them. *See pages 1335–1336.*

CIRCULATORY AND RESPIRATORY SYSTEMS An extremely fast metabolic rate makes huge demands on the animal's heart and lungs. *See pages 1337–1338.*

DIGESTIVE AND EXCRETORY SYSTEMS Because meat is easy to digest, the digestive system is relatively simple. *See pages 1339–1340.*

REPRODUCTIVE SYSTEM Many species have systems that delay the birth of young until prey are easy to find. *See pages 1341–1343.*

External anatomy

COMPARE the body form of a weasel with that of a *RAT*. The rat is a rodent, related to the voles and lemmings that are the weasel's main prey. A rat's slim body is adapted for burrowing in the same way as a weasel, but it is not as long, nor is it as strong for its size.

CONNECTIONS

The smallest of all the true carnivores, the least weasel is a very small animal. At a glance, a female of the small arctic race of least weasel looks barely larger than the voles that she hunts, although her body is actually much longer. This long and thin shape is ideal for a weasel's hunting strategy, which consists of diving down holes and through tunnels in pursuit of burrowing prey. In the far north, least weasels hunt lemmings through the grassy runs that they make beneath the winter snow, insulated from the bitterly cold air and hidden from airborne killers such as the snowy owl.

The least weasel's larger relatives, the ermine and the long-tailed weasel, also hunt in burrows and beneath the snow. Like the least weasel, they have a long, slender, flexible body; short legs; a pointed head; and small rounded ears. All these features are adaptations for hunting in tunnels and runs. In England, the ermine, or

▼ **Least weasel**
The fur is brown above and white below. It turns entirely white in winter except in western Europe and southern Russia. Unlike its close relative the ermine, it has no black in its tail.

The **head** *is flat-topped, with a sharp, almost triangular face. The head is the broadest part of the animal: if the weasel can get its head into a burrow, it can get its body in, too.*

The **ears** *are small and rounded, making movement through burrows easy. The ears are very sensitive.*

The spine is very flexible, allowing the long **body** *to twist and turn through narrow, winding burrows.*

The **eyes** *face forward, but they are well spaced. This arrangement gives the weasel binocular vision— ideal for judging distances when attacking its prey—and also good lateral vision so the weasel can watch for danger.*

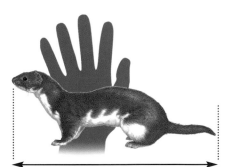

10 inches (25 cm)

Each limb has five sharp, curved **claws**. *Unlike a cat, a weasel cannot retract its claws.*

stoat, pursues mice and voles through old field walls built from loose stones. As it slips sinuously into and out of the cavities between the stones, the ermine resembles a lithe, furry snake—and to its victims it is just as deadly.

Northern weasels

The least weasel, ermine, and long-tailed weasel form a group called the northern or snow weasels. The largest of these species, the long-tailed weasel, lives only in North America, but the two others also occur in northern Asia and in Europe as far south as Spain. Their size varies depending on where they live and their sex: males are bigger than

PREDATOR AND PREY

Trick of the tail

The ermine has a black tip to its tail, which it retains even if the remainder of the animal turns white in winter. The black tip shows up against the snow, apparently ruining the camouflage effect. Experiments have shown, however, that the black tip helps ermines survive attacks by their main enemies, birds of prey. An attacking bird is often distracted by the black tip, so instead of seizing the ermine's body it makes a grab at the tail. Since the tail within the black tuft is very slender, the bird cannot get a grip—and the ermine is able to slip from the hunter's grasp.

*The body is covered with dense fur. The **wavy line** between the brown back and white belly shows that this weasel belongs to a subspecies that does not turn white in winter. In most subspecies, the border between the brown and white is straighter.*

Grison

Least weasel

▲ Grison and least weasel
The grison is the largest of the weasels and polecats. It grows to 28 inches (70 cm) from nose to tail tip, and large individuals weigh 7 pounds (3.2 kg). Least weasels living in the far north may measure only 4 inches (11 cm) and weigh less than 1 ounce (30 g) when mature.

*The **tail** is relatively short and has little function.*

*The **limbs** are short but strong and equipped with sharp, curved claws.*

Snow white

In the north of their range, northern weasels molt their brown fur in winter and grow a thick coat of pure white, for better camouflage in the snow. Farther south, however, weasels of the same species keep their brown fur all year round. The change in fur color is triggered by temperature, but some populations, such as the ermines on Vancouver Island, Canada, and the weasels that live in England, seem genetically incapable of turning white however cold it gets. These weasels have a wavy dividing line between the brown fur of their back and the white fur of their belly. In contrast, "changeable" ermines and weasels have a straight dividing line. The shape of the dividing line and the ability to turn white are probably linked by the same genes.

females. There are also eight other weasels of similar size and shape, including the kolinsky, or Siberian, weasel and the Colombian weasel, which was discovered as recently as 1978.

Weasels share their long, short-legged shape with their nearest relatives, the larger polecats and ferrets. These live mostly on temperate prairie and steppe grasslands, where they eat relatively large prey such as rabbits and

marmots. However, the polecats and ferrets are adapted to catch prey underground in the same way as weasels do. Tame ferrets (domesticated polecats) have long been used to chase rabbits from their warrens, and on the prairies of the American Midwest the very rare black-footed ferret specializes in hunting prairie dogs (ground squirrels) in their burrows.

Luxuriant fur

The weasels and polecats have thick fur, especially in winter, when some species turn white. Some of their relatives—the minks—are famous more for their fur, or pelt, than any other feature. Roughly the same size and shape as polecats but with bushier tails, minks have a luxuriant chocolate-brown pelt composed of long guard hairs that are each surrounded by nine to 24 underfur hairs. The underfur provides insulation against the cold in northern continental winters. The furry insulation is vital because minks and their weasel relatives have very little fat beneath the skin.

An American mink has the same long physique as a polecat or weasel and often pursues prey into burrows in the same way. The American mink's slender, streamlined shape is

▶ Ermine
The ermine, or short-tailed weasel, is larger than the least weasel. The two species have a similar body structure, but the ermine can always be identified by the black tip to its tail.

also ideal for swimming, and this species often hunts in the water like an otter. When submerged, a mink's dense underfur retains an insulating layer of air that keeps the animal warm. Despite not having webbed feet, the American mink is an excellent swimmer. Thus although the mink's body form probably evolved in response to a burrowing lifestyle, it has enabled the animal to exploit a very different habitat.

Tree climbing

Another branch of the weasel subfamily—the martens—is adapted for climbing trees. The martens form a group of nine species that are typically larger than polecats and minks, less long in shape, and with longer legs. Martens have a bushy tail, and like minks they have very dense fur. One species of marten, the sable, which lives in the cold coniferous forests of Russia, has especially dense fur. Martens are extremely agile, and this agility—combined with their sharp claws and excellent balance—enables them to hunt high in trees. The European pine marten, for example, regularly kills squirrels and small birds, as does the smaller American marten.

The sable, European pine marten, and American marten look very similar, with rich brown fur and a pale throat patch, and they live in the same type of northern forest habitat. The American marten shares part of its range with the fisher, the largest of the martens. The fisher

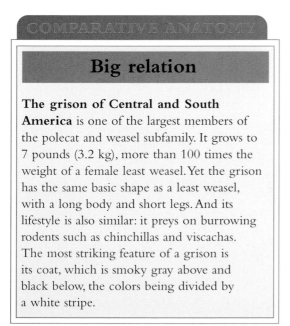

COMPARATIVE ANATOMY

Big relation

The grison of Central and South America is one of the largest members of the polecat and weasel subfamily. It grows to 7 pounds (3.2 kg), more than 100 times the weight of a female least weasel. Yet the grison has the same basic shape as a least weasel, with a long body and short legs. And its lifestyle is also similar: it preys on burrowing rodents such as chinchillas and viscachas. The most striking feature of a grison is its coat, which is smoky gray above and black below, the colors being divided by a white stripe.

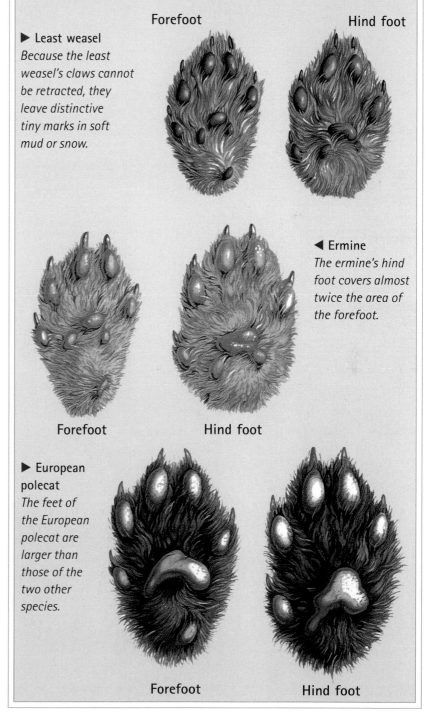

IN FOCUS

Furry feet

Weasels, polecats, and their relatives have five toes on their front and back feet. Each toe is equipped with a strong, sharp, nonretractable claw. The feet generally have furred soles for insulation in cold climates, but a number of leathery pads give grip on rocks and other smooth surfaces. Some species, such as the grison—but not the semiaquatic minks—have partly webbed feet that help them swim more efficiently.

Forefoot   Hind foot

▶ **Least weasel**
Because the least weasel's claws cannot be retracted, they leave distinctive tiny marks in soft mud or snow.

◀ **Ermine**
The ermine's hind foot covers almost twice the area of the forefoot.

Forefoot   Hind foot

▶ **European polecat**
The feet of the European polecat are larger than those of the two other species.

Forefoot   Hind foot

◄ *The wolverine is a large, powerful member of the subfamily Mustelinae. It hunts prey larger than itself, such as reindeer and caribou. Wolverines live in arctic tundras and subarctic coniferous forests in northern North America, Europe, and Asia—all regions that are very cold in winter. The wolverine's large feet spread the animal's weight as it walks on snow.*

and shades of yellow. The African striped weasel is mostly black, with four white stripes and three black stripes running the length of its back. Its tail is mostly black, with a narrow black stripe. Even more dramatic is the zorilla, or African polecat, which has black and white stripes similar to those of a skunk. The conspicuous pattern of a skunk acts as defense, warning its enemies that it is armed with a chemical weapon. Because polecats also have powerful

lacks the pale throat patch but otherwise looks similar and is famous for its ability to kill and eat North American porcupines.

Warning colors

Most of the other species in the weasel subfamily are different enough to be classified in their own genera, although this classification may change as their DNA is studied and compared. The tayra of tropical American forests, for example, is often grouped with the martens because it looks like a long-legged fisher, but it is classified separately because its diet and behavior are different. The tayra may be a marten that has evolved in a different way to suit its tropical habitat. Alternatively, it may have a separate ancestry but has come to resemble the martens because it shares their tree-climbing lifestyle.

The distinctive features of some other species are dazzlingly obvious because their coat is brightly patterned. They include the marbled polecat of central Asia, which has a typically long and thin polecat shape but sports flamboyant spots and stripes of black, white,

▶ *The beech marten is one of nine species of martens. It has a distinctive white bib on otherwise chocolate-brown fur. A large individual may weigh eight times more than a least weasel. Beech martens hunt mostly in forests but sometimes visit farmyards to hunt for rodents.*

scent glands, the bright patterns of these mainly southern species may help protect them in a similar way.

Heavyweight hunter

Although some of the southern weasels and polecats are strikingly colored, their basic anatomy resembles that of their camouflaged northern relatives. The wolverine, however, is different. Occurring all around the arctic region in northern Asia and North America, it is a massively built hunter and scavenger, weighing as much as 55 pounds (25 kg). It is well adapted for its arctic habitat with thick, dark brown to black fur, with paler bands along its flanks and over its eyes. Its broad feet enable it to walk over deep snow without sinking in, giving it an advantage over its main prey, caribou and reindeer. It is essentially a giant marten, but in many ways it looks and behaves like a small bear.

Badgers and otters

The other members of the family Mustelidae are the otters, which are aquatic hunters, and the badgers, which are expert diggers. The American badger, which can weigh up to 26 pounds (12 kg), spends much of its time

digging out burrows for rodents. All species of badgers live in burrows, called dens or sets, and those of the European badger are the most impressive. Some dens have been occupied for hundreds of years, and one was found to consist of 0.5 mile (0.9 km) of tunnels, with 129 separate entrances.

▲ *Minks hunt in terrestrial and aquatic habitats, and their prey reflects this diversity: fish, crabs, and burrowing mammals are all eaten.*

◀ *This ermine is dragging a rabbit back to a safe place to eat. Typically, an ermine hunts larger prey than does its cousin, the least weasel.*

Skeletal system

COMPARE the short jaw and powerful teeth of a weasel with those of a *LION*. For both animals, their teeth are their main weapons, and both sets are adapted for biting with as much force as possible.

According to folklore, the head of a least weasel will fit through a wedding ring. The skull of a below-average size female is certainly small enough. It is tiny in comparison with that of other carnivores and is elongated into an almost tubular form. Despite that, the jaws are relatively heavily built and very strong, able to withstand the stresses exerted by its powerful jaw muscles when it is killing its prey. Weasels regularly kill small animals by punching their sharp canine teeth clean through their victims' skulls. That action demands strong teeth and a tough skull and jawbones to support them.

A weasel's jaws are also able to exert massive pressure because they are relatively short and hinged roughly halfway along the head. The position of the hinge reduces the distance between the jaw muscles and the stabbing teeth at the front of the jaw, so maximizing the power of the muscles when they contract to deliver a lethal bite.

This feature also helps a weasel when it bites through its victims' flesh, using its carnassial teeth. Carnassials are cheek teeth that have evolved into sharp blades instead of flat grinders. The carnassials shear past each other like scissor blades and slice through skin, sinew, and meat in much the same way. Carnassial teeth are the distinctive feature of carnivores (mammals in the order Carnivora) and are most highly developed in dedicated predators like weasels and polecats.

Flexible spine

The bones of a least weasel's skeleton do not need to support much weight, and most of them are very slender. The limb bones are

IN FOCUS

Eaten alive

Weasels are often attacked by a tiny parasitic worm named *Skrjabingylus nasicola*. The worm lives in its victim's head and eats holes in its skull. Between one-quarter and one-half of all least weasel and ermine skulls examined show signs of worm damage. This infestation may explain accounts of strange behavior such as "dancing" in front of their prey. The dance may be a deliberate hunting tactic to baffle their victims, or it may be the writhings of an animal driven mad by pain or suffering from a fit triggered by worms burrowing into its brain.

▼ **Least weasel**
The skull and jaw arrangement is distinctive. It allows the weasel to bite with extreme power. Note also the short, narrow leg bones.

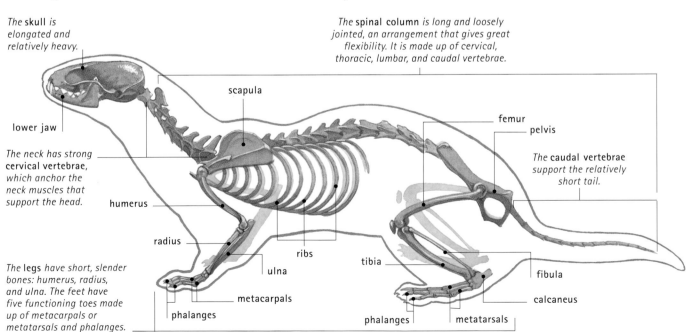

The skull is elongated and relatively heavy.

The spinal column is long and loosely jointed, an arrangement that gives great flexibility. It is made up of cervical, thoracic, lumbar, and caudal vertebrae.

scapula

lower jaw

femur

pelvis

The neck has strong cervical vertebrae, which anchor the neck muscles that support the head.

The caudal vertebrae support the relatively short tail.

humerus

radius

ribs

tibia

fibula

ulna

calcaneus

The legs have short, slender bones: humerus, radius, and ulna. The feet have five functioning toes made up of metacarpals or metatarsals and phalanges.

metacarpals

phalanges

phalanges

metatarsals

COMPARATIVE ANATOMY

Burly badgers

Whereas weasels, polecats, and martens are sleek, agile creatures that are often skilled climbers, their relatives the badgers are stocky, powerful animals that are specialized for digging. Badgers' limbs are much stronger and equipped with long claws for excavating deep burrows. The front claws of the American badger are particularly powerful, and it uses them to dig through the earth in search of burrowing rodents such as ground squirrels and gophers. The other badgers are more general feeders (omnivores), eating insects, berries, and roots as well as small mammals. The European badger feeds mostly on earthworms; its cheek teeth have broad grinding surfaces instead of the slicing carnassial blades of a weasel.

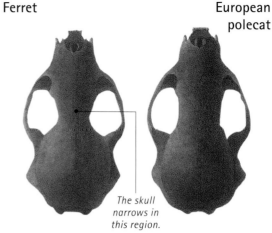

Ferret   European polecat

The skull narrows in this region.

▲ SKULLS FROM ABOVE

There is a slight narrowing of a ferret's skull just behind the eye. This feature is absent in a polecat.

▶ SKULLS

The skulls are long, narrow, and relatively heavy. The lower jaw is short, allowing the teeth to bite with maximum pressure when prey is attacked.

▼ *A weasel's loose-jointed spine allows it to bend its body up and down and from side to side in twisting burrows.*

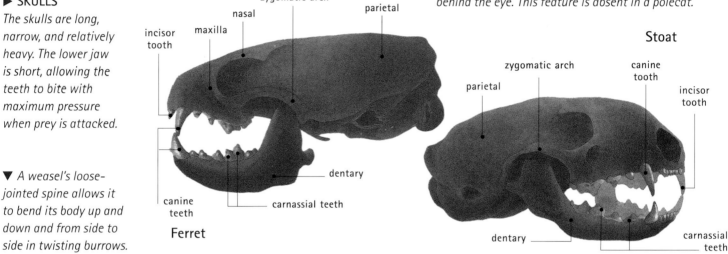

zygomatic arch
nasal
parietal
incisor tooth
maxilla
dentary
canine teeth
carnassial teeth

Ferret

Stoat

parietal
zygomatic arch
canine tooth
incisor tooth
dentary
carnassial teeth

more lightly built than those of similar-size animals with longer legs. This structure works because shorter-legged animals do not suffer so much bending stress. Even the limb bones of larger mustelids such as the American mink are relatively lightweight. A weasel's pelvis is narrow, allowing the animal to squeeze through crevices and burrows, and the shoulder blades are loosely attached to a very small collarbone, allowing maximum flexibility.

The strongest sections of the skeleton are the neck vertebrae, which support the relatively heavy skull. The first two of these vertebrae are larger than the others, with broad flanges to anchor the powerful neck muscles. The other vertebrae are loosely articulated like those of a cat, giving the animal the ability to squirm through twisting tunnels and coil around its prey almost like a snake.

Muscular system

Weasels are small mammals, but they are also powerful. They can kill animals that are much bigger than themselves, partly because they have a very efficient set of skeletal muscles: the muscles that give them the power of movement. Muscles are masses of fibers made from two types of proteins that lie side by side in long filaments. When a muscle receives a signal from the nervous system, the two filaments ratchet past each other, making the overall fiber shorter. On a larger scale, this makes the muscle contract and exerts a considerable force.

Skeletal muscle

Skeletal muscles are arranged in sets that work against each other. One set of muscles flexes a leg, for example, and another set straightens it. The arrangement is necessary because muscles can exert power only by contracting. When a muscle relaxes, it has to be extended by another muscle working in opposition. This effect is most clearly seen in the weasel's strong back muscles, which allow it to coil around its prey as does a snake. If the muscles on the left of its spine contract, the spine bends to the left. If these muscles then relax while those on the right contract, the spine bends to the right.

Some of a weasel's strongest muscles are the large temporalis muscles, which are attached to the upper flange of the jawbone and anchored to the ridge on top of the skull. These muscles provide the jaw with power that the weasel needs to kill its prey and carry them to a safe refuge to feed its young. In contrast, the leg muscles, including the biceps, triceps, extensor carpi digitalis, flexor carpi ulnaris, and gastrocnemis, are relatively small.

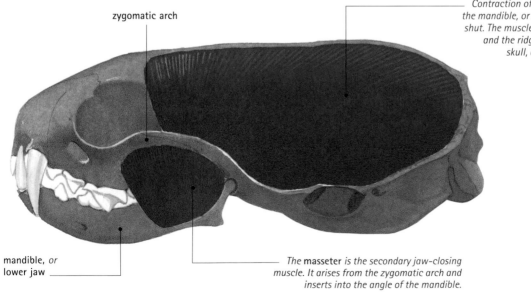

zygomatic arch

Contraction of the temporalis muscle elevates the mandible, or lower jaw, and snaps the mouth shut. The muscle arises from the zygomatic arch and the ridge that runs along the top of the skull, and it inserts into the lower jaw.

mandible, *or* lower jaw

The masseter is the secondary jaw-closing muscle. It arises from the zygomatic arch and inserts into the angle of the mandible.

◀ JAW–CLOSING MUSCLES
Least weasel
The temporalis and masseter are two of the largest sets of muscles in a weasel's body.

Nervous system

Like most predators, a weasel is an intelligent animal with highly developed sensory systems. It needs acute senses to gather clues to the whereabouts of possible prey, and it needs the mental power to devise a hunting strategy based on that information. As it patrols its territory, a weasel uses all its senses to gather information. It has keen eyesight, an excellent sense of smell, and sharp hearing.

Acute hearing

A weasel's sense of hearing is very sensitive, with a resonating chamber in each middle ear that helps compensate for the small size of its outer ears. A weasel is very sensitive to high-pitched sounds such as the squeaks of voles and other small rodents. It can also hear the quiet rustles of a mouse walking through grass. By swiveling its head until the sound is equally loud in both ears, the weasel can pinpoint its target even in total darkness.

A weasel can also use its sense of smell to follow the tracks of prey and to locate and identify other weasels. Scent is particularly valuable in hunting underground or beneath the snow. A weasel can distinguish the fresh scent of a fleeing vole from the stale scent inside a vole's burrow. Thus the weasel can avoid taking a wrong turn. Scavenging

Scent marking

Most weasels are basically solitary animals that communicate almost exclusively by scent. As they move around their territory they leave scent signals in the form of urine and feces perfumed by their anal scent glands. They also have scent glands on their neck, chin, and feet. Some species are more strongly scented than others: the European polecat, for example, was once called the "foulmart" because of its strong smell. Socially dominant individuals also produce stronger scents, and simply sniffing at a scent mark enables a weasel, polecat, or mink to assess the marker's status, and even its identity.

wolverines rely on their sense of smell to lead them to the carcasses of dead animals buried beneath the snow, and mink depend upon scent when hunting on land. In water, mink probably rely upon their touch-sensitive whiskers, since their nostrils are clamped shut when submerged, and their eyes are not adapted for seeing underwater.

▼ **Least weasel**
As with other predators, a weasel's senses of vision, hearing, and smell need to be very sensitive. Information is carried to the brain from the eyes along the optic nerves; from the ears along the auditory nerves; and from the nose along the olfactory nerve.

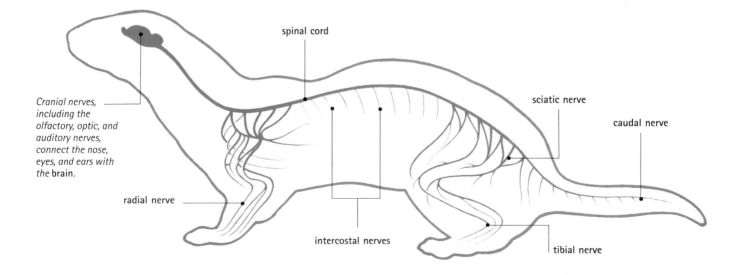

Cranial nerves, including the olfactory, optic, and auditory nerves, connect the nose, eyes, and ears with the brain.

spinal cord

sciatic nerve

caudal nerve

radial nerve

intercostal nerves

tibial nerve

COMPARE the eyes
of an American
mink with those
of an *OTTER*. They
both hunt in the
water, but the
otter's eye contains
powerful muscles
that change the
shape of the lens
for improved
underwater vision.
As a result, a diving
otter can see three
times better than
a mink can.

CONNECTIONS

▶ *Excellent senses of
hearing, smell, and
sight make European
polecats efficient
hunters. They hunt
mostly at night for a
wide range of prey,
especially rabbits.*

Information processor

A weasel's senses are part of its peripheral
nervous system: a network of fibers called
neurons that carry electrical signals between all
the organs and muscles of its body. Its senses
generate signals that pass through sensory
neurons to the spinal cord and brain that form
its central nervous system (CNS). Interneurons
within the CNS process the information
gathered by the senses and trigger signals that
pass back out, through separate motor neurons,
to the muscles and other effector organs.

Some of these actions are automatic—for
example, breathing and the reflexes that make
the weasel recoil from pain. These actions are
controlled by interneurons in the brain stem and
spinal cord. Conscious actions are controlled
by the largest part of the brain, the cerebrum,
which processes incoming information and
compares it with data that have been stored as
memory. Picking up the scent of a vole, for
example, is of little use if the weasel has
no mental record to compare the scent with.
Some kinds of sensory information, of course,
demand less processing than others, and to a
weasel the scent of a vole triggers a reaction
that is almost as quick as a reflex.

PREDATOR AND PREY

Smelly defense

Some weasels and their relatives have
developed the ability to defend themselves
with an acrid, sulfurous spray from their
anal glands. The masters of this tactic are
the skunks, which belong to a different
but closely related family, the Mephitidae.
However, the very skunklike zorilla can
fend off attack in the same way, and so can
the marbled polecat, among others. The
flamboyant coat patterns of all these species
advertise their pungency and often warn
off their enemies before the spray becomes
necessary. This also helps conserve supplies
of the scent, which costs energy to produce.

Circulatory and respiratory systems

A weasel is a highly specialized hunter, suited to pursuing small animals through runs and burrows. It shares this characteristic with many of its relatives, including the ermine, the black-footed ferret, and the European polecat. They all have the same long, thin body suited to squeezing through narrow tunnels. This shape has a disadvantage, however: it gives the animals a large surface area compared with their volume, so they lose heat easily. Least weasels, in particular, live as far north as arctic regions, where they use a huge amount of energy simply keeping warm. Least weasels and other species that have to endure cold conditions therefore need highly efficient respiratory and circulatory systems to keep the energy flowing.

▼ **Least weasel**
Inhaled oxygen reaches a weasel's bloodstream by way of the lungs. Oxygen-rich blood from the lungs travels to the heart, which pumps the blood around the rest of the body.

Fuel supply

Like all animals, a weasel requires energy to power its body processes and muscles. It converts the energy from food into the blood sugar glucose. Sugars store energy in the form of chemical bonds, which act like wire ties around tightly coiled springs. When the bonds are broken, the energy is released. The process that breaks the bonds is a chemical reaction

called oxidation, which involves mixing the blood sugar with oxygen. In weasels and other mammals the oxygen and sugar that fuel the reaction are both delivered in the blood.

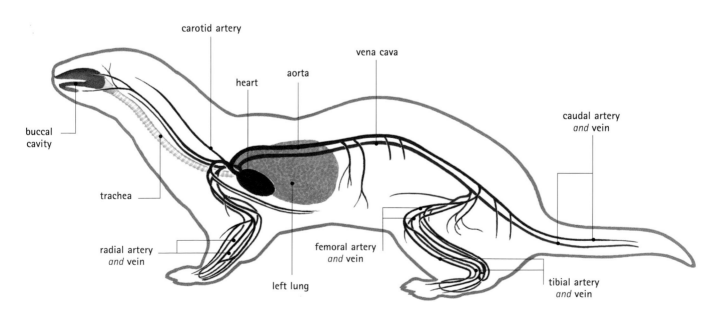

carotid artery

vena cava

heart   aorta

buccal
cavity

caudal artery
and vein

trachea

radial artery
and vein

femoral artery
and vein

tibial artery
and vein

left lung

Oxygen is gathered from the air by the weasel's two lungs, which are made up of tiny, delicate air sacs, or alveoli, connected to the animal's windpipe, or trachea, by a branching network of tubes. The lungs are contained in an airtight chest cavity, which is sealed at the bottom by a sheet of muscle called the diaphragm. When the diaphragm contracts, the volume of the chest cavity increases, and the lungs expand, causing the animal to breathe in. Air is drawn into the trachea and alveoli. Oxygen in the air passes into thin-walled blood vessels around the alveoli and bonds to the pigment hemoglobin in red blood cells.

Thick-walled arteries

The oxygen-rich blood returns to the left side of the heart, which pumps it through a network of thick-walled arteries, and then increasingly small blood vessels called capillaries, to all the tissues of the body. As it flows, the blood picks up glucose from the digestive system and liver; and when the blood reaches the cells, the glucose combines with the oxygen to release its energy. The reaction turns the glucose and oxygen into carbon dioxide and water, which are carried back to the right side of the heart in blood flowing through the veins. The heart then pumps the blood back through the lungs, where the waste carbon dioxide and water pass out into the air and are replaced by more oxygen.

Tireless muscle

A weasel needs a steady supply of glucose and oxygen to keep active, and its brain in particular cannot survive even a few seconds without oxygen. Its heart is made of a type of muscle called cardiac muscle, which does not become exhausted. Thus the heart is able to beat continuously. Even when the animal is at rest, its heart beats about 360 times a minute; when it is active the rate is even faster. As with other vertebrates, the heart muscle is fueled by glucose and oxygen, delivered through a network of coronary arteries. If these arteries become blocked for any reason, the animal suffers a heart attack and dies.

◀ Despite its protective burrow and thick fur, a least weasel loses heat very quickly in the winter snow. To compensate for the heat loss, it needs an efficient circulatory system and an almost constant supply of food.

Digestive and excretory systems

CONNECTIONS

COMPARE the relatively simple digestive tract of a least weasel with the complex, multistage digestive system of a **RED DEER**. The deer feeds on grass and leaves, which are much harder to digest than meat, and its intestine has features to convert tough plant fiber into useful food.

Weasels, minks, and polecats are dedicated predators. Except for an occasional feast of soft, sugar-rich berries or other fruits in late summer, they eat only meat. Meat is easy to digest, and these animals' digestive system is relatively simple. Weasels do not need special adaptations to cope with tough plant fiber and turn it into usable food. They do not even need to chew their food, because unlike plant cells the cells of soft animal tissues do not have to be crushed to allow access to digestive juices. A weasel can eat its meal very quickly by swallowing food in big chunks, without risking indigestion.

Accordingly, a weasel's teeth are suited to killing animals and slicing meat rather than chewing. A weasel has a relatively large, simple stomach that can hold a lot of meat. However, since the meat is digested relatively quickly, a weasel needs only a short intestinal tract. Like all carnivores a weasel is adapted for eating big meals at irregular intervals—whenever it can catch a victim—rather than for the steady food intake of a typical herbivore.

Molecular breakdown

The digestive juices secreted by the walls of a weasel's stomach and small intestine contain substances called enzymes. These are proteins

COMPARATIVE ANATOMY

Omnivorous relatives

The smaller weasels of the genus *Mustela* are almost exclusively predatory, eating virtually nothing but the meat of small mammals and birds. Some of their larger relatives have a broader diet, however. The Eurasian pine marten feeds mainly on birds and mammals, but it also eats earthworms, insects, berries, wild apples, nuts, and fungi, especially when these are seasonally abundant. The South American tayra, which resembles a marten, eats a lot of fruit. Eurasian badgers feed mainly on earthworms plus insects, berries, and even roots. The larger size of pine martens and tayras increases the size of their digestive tract and reduces their need for instant energy, allowing them to eat and digest a wider range of foods.

that break the chemical bonds binding complex protein molecules together, reducing them to simpler molecules called amino acids. The amino-acid molecules are the building blocks of all proteins, so a weasel's body can use them

▼ **Least weasel**
Weasels have a short digestive tract with a simple stomach, suited to a diet of meat (which is easy to digest).

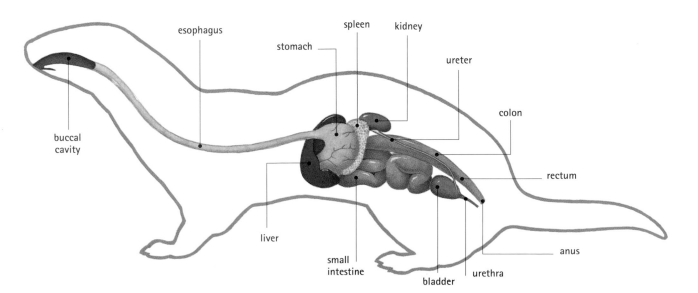

esophagus · spleen · kidney · stomach · ureter · buccal cavity · colon · rectum · liver · small intestine · bladder · urethra · anus

to make the proteins that it needs. Meanwhile, other enzymes get to work on carbohydrates and fats, reducing them to simpler compounds that the weasel can turn into energy.

All the products of digestion are absorbed through the wall of the small intestine into the bloodstream, which then passes through the liver. This organ refines the products of digestion further, transforming them into substances that the weasel's body can use as fuel or building material to make body tissues such as muscle and bone. The liver also neutralizes any mild poisons in the diet. The waste products of these processes are turned into harmless substances that are carried in the bloodstream to the kidneys. There, the wastes are filtered out of the blood, mixed with water, and excreted as urine. Meanwhile, any indigestible material, including bones and hair, passes into the large intestine. There, the water in the waste is removed, so that the waste is compacted into pellets that are voided as feces.

EVOLUTION

Size matters

All northern weasels hunt in much the same way, but they take different prey according to their size. In Europe, least weasels feed mostly on small voles and mice, whereas larger ermines eat water voles and rabbits. This prey selection allows both species to live in the same habitats without competing directly for food. Where one species is absent, however, the other takes over its role. In Spain, where there are no ermines, weasels often grow to the size of ermines. In Ireland, where there are no least weasels, the ermines are usually the size of weasels. Thus it seems that each species evolves differently, depending on the competition for prey.

▲ *Small mammals, including red squirrels, form an important part of a pine marten's diet. Pine martens use their tail to aid balance as they chase the squirrels along the branches of trees. Pine martens are omnivores, however, and berries may make up 30 percent of their summer diet.*

Reproductive system

All the weasels, polecats, and martens and their relatives in the subfamily Mustelinae are usually solitary animals. The males and females come together only to mate, and the males take no part in rearing the young they have fathered. Each male claims a territory and attempts to mate with any females that pass through his patch. Rival males compete for territories, and because stronger, healthier males are more successful at this, they mate more frequently.

Females often mate with several males, and in species such as the American mink females may produce litters with more than one father. The last male to mate usually fathers most of the litter, however, and since this is usually the strongest male (who stops weaker males from taking his place) this may ensure that the fittest male leaves more descendants.

Rival males become very aggressive toward one another during the mating season, often fighting over territory. The territorial success of a male marbled polecat is actually visible, since high levels of the hormone testosterone in his blood make his fur slowly change color from black and buff to black and orange. The more successful the male is, the more extra testosterone is produced and the more orange his fur becomes. A vividly colored male will usually send off a duller one, but if the brighter animal is beaten by a rival his orange glow gradually fades to buff again.

COMPARE the tiny, naked, helpless newborn of a weasel with the highly developed calf of a *WILDEBEEST*. The wildebeest calf is able to run after its mother within just 10 minutes of being born. Thus it stands a chance of escaping from the predatory animals that prowl the African plains.

CONNECTIONS

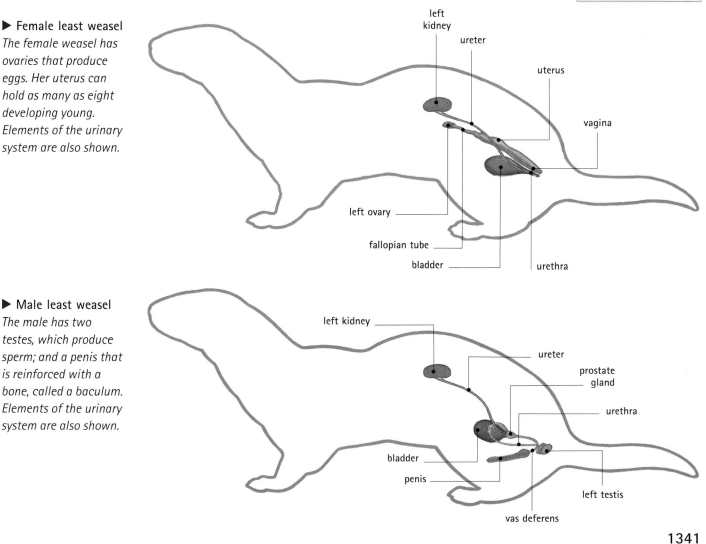

▶ **Female least weasel**
The female weasel has ovaries that produce eggs. Her uterus can hold as many as eight developing young. Elements of the urinary system are also shown.

left kidney
ureter
uterus
vagina
left ovary
fallopian tube
bladder
urethra

▶ **Male least weasel**
The male has two testes, which produce sperm; and a penis that is reinforced with a bone, called a baculum. Elements of the urinary system are also shown.

left kidney
ureter
prostate gland
urethra
bladder
penis
left testis
vas deferens

▶ *European ferret kittens' eyes open after about four weeks. There are usually between 10 and 15 kittens in a single litter.*

Delayed implantation

Mating may be repeated several times during a period of up to three hours. The male's penis is reinforced by a bone called the baculum, and this bone helps keep the penis erect during long mating sessions. The vigor of the mating act is important because it stimulates the female's reproductive organs into producing eggs. This response is called induced ovulation. The eggs are then fertilized by the male's sperm and become fertile eggs, or zygotes. These implant into the wall of the uterus and develop into embryos.

In some species, such as the ermine and the long-tailed weasel, the implantation of the fertilized eggs is delayed by several months, extending the pregnancy to ensure that the young are born when food is most plentiful. Ermines, for example, mate in summer. The true gestation period is about one month, and if the young developed without delay they would be born in the fall. They would then probably fail to survive the winter, especially in the far north. However, implantation is delayed by nine months, and the young are born in spring when there are plenty of prey, and a warm summer lies ahead. Minks and martens also delay implantation by 2 to 10 months.

Boom and bust

The least weasel does not practice delayed implantation. That is surprising, since this animal is closely related to the ermine and lives in the same northern regions with their harsh winters. Least weasel eggs implant within 12 days of being fertilized, and the young are born just 25 days later. A female is sexually mature at the age of just three months, so if she is born in April she is able to produce her own litter of four to six young in August. That time of birth gives the young just long enough to grow to a good size before winter closes in. The following year a female least weasel may have one litter in spring and a second litter in summer, something that the larger, slower-growing ermine cannot manage.

The least weasel's ability to breed rapidly is partly a consequence of its small size and fast metabolism. Its life races past at such a rate that it matures young and may not live much longer than a year, so it must breed as quickly as possible. Fast breeding is also a response to the unpredictable food supply in the far north. Its main prey in the arctic, rodents called lemmings, are famous for their population explosions in good breeding seasons. The least weasel's ability to breed fast enables it to make the most of these boom years, and at such times it may even breed in winter, beneath the snow. When the lemming population crashes,

COMPARATIVE ANATOMY

Big brother

Male weasels are always bigger than their sisters. Male fishers (large martens) and least weasels are often twice as heavy as the respective females. No one knows the reason for this weight difference. It may help reduce competition for prey between the sexes because the smaller females target smaller animals. The difference may reduce the female's food needs, allowing her to supply more food to her young. Or the difference may simply be the result of competition between males that favors bigger, stronger individuals who then pass on their genes to their more numerous descendants.

many weasels die, too, but enough survive to multiply rapidly when the lemming population builds up again.

The ermine has a different method of responding to the boom-and-bust prey supply in the far north. It has twice as many young in each litter, with up to 12 being common. If there are plenty of prey, most of the young survive. If not, many may be born dead. Because the young are very small when they are born, this does not represent a huge waste of the mother's resources.

Male ermines are known to mate with infant females in the nest, despite being the current territory holder and, possibly, their father. Even at less than a month old the nestlings produce eggs that can be fertilized, but they do not give birth until the following spring. This behavior ensures that females breed as young as possible but it also increases the risk of serious inbreeding and genetic disorders. However, the short life span and 10-month pregnancy of ermines make it likely that the nestlings' real father is usually dead, and the male that mates with them in the nest is probably no relation.

By contrast, species such as martens, which live in more stable forest habitats, have smaller litters but a much lower infant death rate.

Helpless young

Young weasels are born blind, deaf, and naked, in a nest that their mother has usually taken over from one of her victims and sometimes even lined with their fur. Newborn weasels open their eyes at three to four weeks old, by

CLOSE-UP

Scar tissue

A mature female American mink often has a small patch of white fur on the back of the neck. The white fur grows on damaged skin, scarred by the sharp teeth of the male that seizes her by the scruff of the neck during mating. The male nearly always draws blood, but because he is bigger and stronger than the female, she can do little to stop him.

which time they are eating meat brought to the nest by the mother. They are fully weaned off milk at six to eight weeks of age, and by that time they are able to kill their own prey, although the family stays together for some weeks after weaning. Larger species of mustelids such as ermines, polecats, minks, and martens develop at a much more slower pace. Young pine martens, for example, cannot kill their own prey until they are at least 10 weeks old, and they do not reach full independence until the age of about six months.

JOHN WOODWARD

FURTHER READING AND RESEARCH

Macdonald, David (ed.). 2006. *The Encyclopedia of Mammals*. Facts On File: New York.

The Hall of Mammals:
www.ucmp.berkeley.edu/mammal/carnivora/carnivoramm.html

◄ At one week old, as here, European polecat kittens are still blind. By four weeks, they are weaned off milk and onto meat. By six months, they are sexually mature.

Weevil

KINGDOM: Animalia CLASS: Insecta ORDER: Coleoptera
FAMILY: Curculionidae

Weevils are beetles, which make up the order Coleoptera. There are more known species of beetles than any other type of insect. There are nearly 1 million known species of insects in the world, and about 7,000 new species are described each year. Of the known species, far more than one-third of a million are types of beetles, and it is very likely that there are 5 million to 10 million species of beetles as yet undiscovered. However, because most of these undescribed species live in rain forests—and taking into account the rate of destruction of the forests—most of these species might become extinct before they are discovered.

Anatomy and taxonomy

Ancestral beetles probably first appeared about 270 million years ago, in the Permian period. These ancestors were probably related to the alder flies (order Megaloptera) because they have similar patterns of veins in the wings, particularly in the elytra (wing cases). These ancient beetles probably lived under bark or in similar places, where they fed on the fungi growing there.

- **Animals** All animals are multicellular life-forms. They cannot make their own food, unlike plants, which produce food by photosynthesis. Instead, animals consume other organisms. Most animals can move around, so they need to respond quickly to information (stimuli) gathered from their surroundings.

- **Arthropods** These animals do not have a backbone—they are invertebrates—but they do have a double, ventral (toward the lower surface) nerve cord. Support is provided by an external skeleton, or exoskeleton, which is a tough, largely rigid layer around the outside of the body. The body is divided into a series of repeated rings, or segments. In more recently evolved arthropods the segments show a variety of modifications, resulting in distinct body regions.

▼ *Weevils make up the family Curculionidae, which is in the order Coleoptera, or the beetles. Beetles are insects—invertebrate animals with a tough exoskeleton. At present, there are about 41,000 known species of weevils. Well-known species include the grain weevil, the cotton-boll weevil, and the nut weevil.*

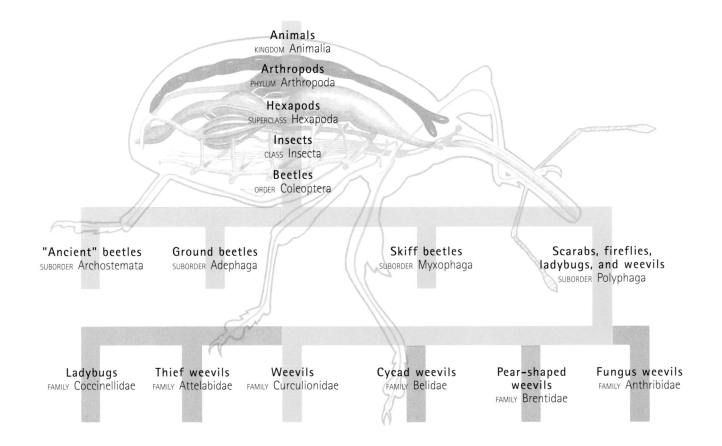

Animals
KINGDOM Animalia

Arthropods
PHYLUM Arthropoda

Hexapods
SUPERCLASS Hexapoda

Insects
CLASS Insecta

Beetles
ORDER Coleoptera

"Ancient" beetles
SUBORDER Archostemata

Ground beetles
SUBORDER Adephaga

Skiff beetles
SUBORDER Myxophaga

Scarabs, fireflies, ladybugs, and weevils
SUBORDER Polyphaga

Ladybugs
FAMILY Coccinellidae

Thief weevils
FAMILY Attelabidae

Weevils
FAMILY Curculionidae

Cycad weevils
FAMILY Belidae

Pear-shaped weevils
FAMILY Brentidae

Fungus weevils
FAMILY Anthribidae

● **Insects** The body of an adult insect is divided into three distinct regions: the head, thorax, and abdomen. The head contains the mouthparts and sense organs such as the eyes for vision and the antennae for smell. The thorax bears the two pairs of wings and the three pairs of legs (the six legs give insects and their very close relatives the alternative name, Hexapoda, which means "six legs"). The third region, the abdomen, contains organs for reproduction, digestion, excretion, and circulation.

● **Coleopterans** At present, the order Coleoptera is split into four suborders containing a total of 168 families—although beetle classification often changes. The suborder Archostemata contains three or four families. These are the most primitive beetles, and a number are known only from fossils. The suborder Adephaga contains 11 families. Most beetles in Adephaga are carnivorous and very active. The Adephaga includes the family Carabidae (the ground beetles). The suborder Myxophaga has four families. These beetles feed on algae. A group of 19 superfamilies with 149 families

▲ *Weevils have a long, downward-curved snout with antennae, as can be seen in this male* Rhinastus latesternus, *which lives in the rain forests of Peru, South America.*

EXTERNAL ANATOMY Beetles have a tough, segmented exoskeleton. The segments are specialized, giving three body regions: the head, thorax, and abdomen. There are a pair of antennae and a pair of compound eyes on the head. Three pairs of legs and usually two pairs of wings arise from the thorax. *See pages 1346–1351.*

INTERNAL ANATOMY Muscles move the legs, mouthparts, and, in flying beetles, wings. A beetle's body also contains the organs and structures of the nervous, circulatory, respiratory, digestive, and reproductive systems. *See pages 1352–1353.*

NERVOUS SYSTEM In the head, fused ganglia form a brain. A pair of ventral nerve cords with ganglia run down the length of the body. *See page 1354.*

CIRCULATORY AND RESPIRATORY SYSTEMS A dorsal contracting vessel—often referred to as the heart—pumps bloodlike hemolymph through vessels and into a body cavity called the hemocoel. Respiration is by spiracles and tracheae, as in other insects. *See pages 1355–1356.*

REPRODUCTIVE SYSTEM The male has sperm-producing testes and a penis, or aedeagus. Females have two egg-producing ovaries and a vagina. Fertilization is internal. *See pages 1357–1359.*

make up suborder Polyphaga. This varied group has species that feed on fungi; plants, including wood; or other animals. Suborder Polyphaga includes the ladybugs (family Coccinellidae) and weevils (family Curculionidae).

Beetles undergo complete metamorphosis, so there are four stages in their life cycle: egg, larva, pupa, and adult. The larval stage is sometimes called a maggot (in flies), a caterpillar (in butterflies and moths), or a grub; the pupa of a butterfly is called a chrysalis, and the adult stage is sometimes called the imago. The larva is the feeding and growing stage. During pupation, the insect may appear to be resting, but actually its body cells are being completely rearranged into the final form—the adult, which is the reproductive stage. Beetles are endopterygote; this term means that their wings develop inside the body during pupation rather than outside (exopterygote) as in earwigs, crickets, and dragonflies.

● **Curculionidae** The members of this family of beetles are commonly called weevils. (*Weevil* is derived from the Old English word for beetle.) There are about 41,000 species in the family Curculionidae, about 2,700 of which are found in the United States. They can be recognized by the presence of a snout, or rostrum, on the front of the head. The mouthparts are situated at the end of the snout. Nearly all weevils feed on plants, and many species are economically significant crop pests. These include the rice weevil and the cotton-boll weevil.

External anatomy

COMPARE the external anatomy of a weevil with that of another arthropod, a *TARANTULA*. A tarantula, like a weevil, has an exoskeleton with jointed legs. However, it differs from a weevil in having eight legs, fangs, a fused head and thorax, and usually six to eight eyes, and it does not have wings.

Beetles are small to large insects, ranging from 0.01 to 7.9 inches (0.001–20 cm) long. Most weevils are relatively small beetles, but they are very variable in size. The largest are the tropical palm weevils. The bodies of these species can measure up to 2 inches (5 cm) long. By contrast, some of the flea weevils and other species that live in soil and leaf litter are tiny, just 0.04 inch (1 mm) long.

Like beetles, weevils occur in almost all terrestrial and most freshwater habitats. No species are known to live in seawater, but some occur on ocean shores. Weevils are one of the few types of animals to inhabit frozen Antarctic islands, although they do not live on the Antarctic continent. They live on every other continent, and the greatest diversity of species is in the tropics.

▶ **Cotton-boll weevil**
The body shape of the cotton-boll weevil resembles a long teardrop. The most characteristic feature of a weevil is the long, downward-curved snout with antennae. Like all other insects, weevils have six legs. The flight wings are kept folded up under the elytra (wing cases).

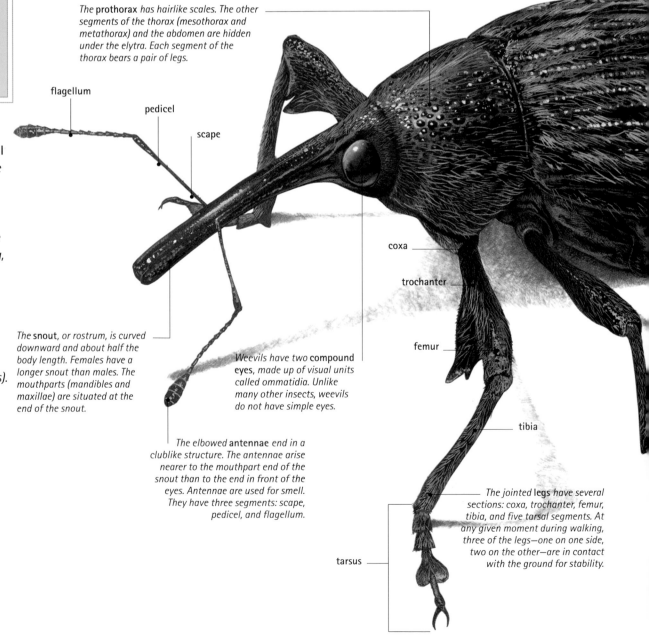

The **prothorax** has hairlike scales. The other segments of the thorax (mesothorax and metathorax) and the abdomen are hidden under the elytra. Each segment of the thorax bears a pair of legs.

flagellum

pedicel

scape

The **snout**, or rostrum, is curved downward and about half the body length. Females have a longer snout than males. The mouthparts (mandibles and maxillae) are situated at the end of the snout.

Weevils have two **compound eyes**, made up of visual units called ommatidia. Unlike many other insects, weevils do not have simple eyes.

The elbowed **antennae** end in a clublike structure. The antennae arise nearer to the mouthpart end of the snout than to the end in front of the eyes. Antennae are used for smell. They have three segments: scape, pedicel, and flagellum.

coxa

trochanter

femur

tibia

The jointed **legs** have several sections: coxa, trochanter, femur, tibia, and five tarsal segments. At any given moment during walking, three of the legs—one on one side, two on the other—are in contact with the ground for stability.

tarsus

Most features of a weevil's body are those typical of all beetles. The adults usually have two pairs of wings, although in some species the hind pair is missing. The scientific name of beetles, Coleoptera, refers to their forewings: *coleo* means "sheath" and *pteron* means "wing." The hardened forewings cover and protect the membranous hind wings, which—in many species, but not all—are used for flight.

Together, these hardened forewings are called the elytra (singular, elytron), and they meet together in the middle at the suture. The elytra are the most obvious feature of a beetle. They have many different forms, or

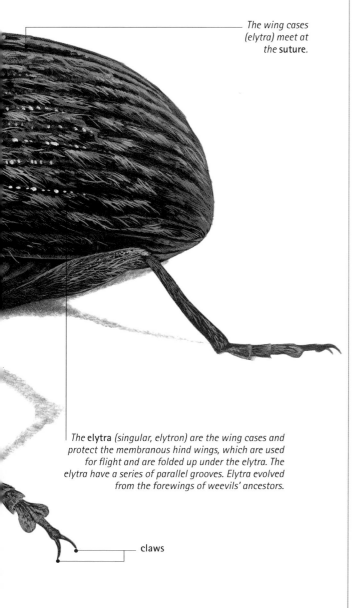

The wing cases (elytra) meet at the suture.

*The **elytra** (singular, elytron) are the wing cases and protect the membranous hind wings, which are used for flight and are folded up under the elytra. The elytra have a series of parallel grooves. Elytra evolved from the forewings of weevils' ancestors.*

claws

Beetle antennae

The antennae are jointed, mobile structures in front of, or lying between, the eyes. Usually, they are much shorter than half the length of the body and have 11 segments (sometimes called flagellomeres), although this number may vary between 1 and 27. There are differences in length, segment number, and form, depending upon the type of beetle.

The scape is the first, or basal, joint of the antenna; it is often longer than the other antennal joints. The pedicel is the second joint, and the flagellum forms the rest of the antenna. Sometimes, the end joints of the antenna are swollen into a club. The part of the antenna between the pedicel and the base of the club is called the funicle. Different families of beetles have different types of antennae.

▼ **COMMON TYPES OF ANTENNAE**

The more obvious antennae of beetles include moniliform, in which the round segments make the antenna look like a string of beads; clavate, in which the segments are wider toward the tip of the antenna, forming a club; and geniculate, in which there is an abrupt bend or elbow part of the way along the antenna—a characteristic of weevil antennae.

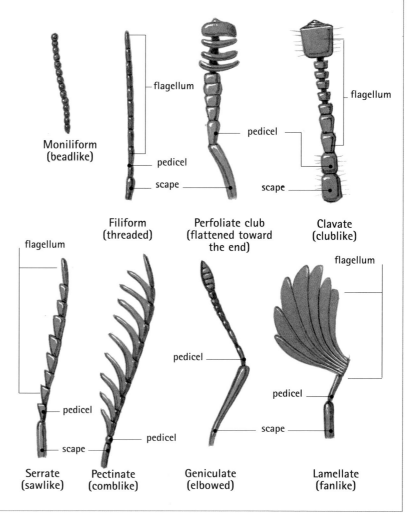

Moniliform (beadlike)

Filiform (threaded)

Perfoliate club (flattened toward the end)

Clavate (clublike)

flagellum — pedicel — scape

flagellum — pedicel — scape

Serrate (sawlike)

Pectinate (comblike)

Geniculate (elbowed)

Lamellate (fanlike)

flagellum — pedicel — scape

1347

Fossils and beetle evolution

Insects appeared in the Devonian period, about 400 million to 360 million years ago. They evolved rapidly, and the first winged insects appeared in the late Carboniferous period, about 300 million years ago. The beetles arose in the early Permian, about 270 million years ago. There are few fossils of beetles from this period, most being of a single elytron. The major beetle suborders evolved during the Triassic period, 248 million to 208 million years ago.

Early ancestral beetles probably lived under the bark of trees, where they fed on fungi, wood that had been attacked by fungi, or other invertebrates. There, their large membranous wings would have been liable to damage and a considerable disadvantage. The forewings evolved into thick, hard structures, forming a protective covering for the delicate hind wings, which were used for flight.

Most beetle families arose during the Jurassic period, and most of the modern beetle families were present by the end of the Jurassic, around 144 million years ago. True weevils were among the last of the major beetle families to appear, in the middle Cretaceous period around 100 million years ago. The oldest known fossil of a true weevil is probably *Curculionites hylobioides*. It was found in the Fox Hill Beds of South Dakota. After that, the number of true weevil fossils increased during the Tertiary period, and weevils are now the most numerous of the various families of beetles.

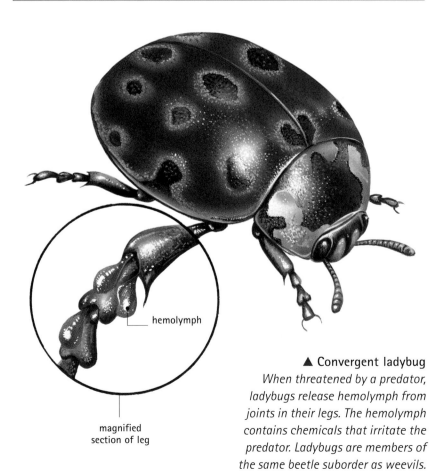

hemolymph

magnified
section of leg

▲ **Convergent ladybug**
When threatened by a predator, ladybugs release hemolymph from joints in their legs. The hemolymph contains chemicals that irritate the predator. Ladybugs are members of the same beetle suborder as weevils.

variations. Often, they have longitudinal raised lines called striae—as in the cotton-boll weevil. In many species of beetles, the elytra have a range of patterns, called sculpturing, in the surface. There may be a number of characteristic dots, or pits, excavated into the surface. Sometimes, each pit has a bristle. The elytra show a range of colors, and some are covered with or have patterns of pubescence ("hairs" made of a tough substance called chitin rather than keratin as in mammals).

Some beetles—including many weevils—have lost the power of flight, and their hind wings have become much smaller. In these species, the two elytra are fused along the suture. In the rove beetles (family Staphylinidae), the elytra are much shorter than the abdomen. Many rove beetles are very active, with a flexible body, and are good fliers. The hind wings are folded several times to fit under the elytra.

Head and mouthparts

As in all other insects, the body of a beetle is divided into three regions: head, thorax, and abdomen. The head is prognathous—its axis is aligned with the long axis of the body, often at a slight downward angle.

Weevils and most other beetles have a pair of compound eyes, one eye on each side of the head, and although beetle vision is good, it is probably not as acute as that of dragonflies. Some cave-dwelling beetles do not have eyes. The males of glowworms and fireflies have large eyes. Only a small minority of beetle species have simple eyes, or ocelli.

Weevils have a snout, or rostrum, at the front end of the head. The structure of the snout is very variable. In some species it is short and broad, and in others it is long and needlelike, reaching up to twice the length of the remainder of the body. A beetle's mouthparts are almost at the tip of the head, which in weevils means at the tip of the snout. The mouthparts are used for biting or chewing, so the mandibles are well developed and move from side to side, except in the nut weevils, in which the mandibles move up and down. Muscles stretch from the head along the length of the snout to work the mandibles at the tip.

Beetles have antennae, and those of weevils are situated on the snout. Weevil antennae are

◄ *The chestnut weevil has a very long snout. The female uses her snout to bore holes in chestnuts or acorns. She then deposits an egg in each hole.*

IN FOCUS

Bioluminescence

Light produced by animals is called bioluminescence and is created in a similar way in different species. Some beetles, including some members of the same suborder as weevils, are bioluminescent. A substance called luciferin is chemically changed (oxidized) by an enzyme called luciferase. This reaction occurs in the presence of molecular oxygen, with energy from ATP and magnesium ions as cofactors. The reaction results in the production of a recyclable product, carbon dioxide; and a photon (unit of light). The product is then chemically recycled back into luciferin and used again. The photon produces the color—each species produces its own characteristic color. This often has a greenish hue but may be red or yellowish green. The color differences depend upon the energy of the photon that is released. These different colors are believed to be a result of variations in the structure of the luciferase enzyme. The chemical process is very efficient: usually 90 percent or more of the released energy appears as light. Therefore, very little energy is lost as heat. For this reason, the light produced is often called a "cold light."

The luciferin–luciferase principle of bioluminescence has important uses in biotechnology and medicine, particularly in tests in which the luciferin–luciferase reaction can be utilized to operate only if a specific substance being tested for is present. Uses include testing the quality of drugs; the rapid screening of drugs against disease-causing organisms; testing for infections such as tuberculosis and HIV; testing for the viability (living potential) of cells, such as sperm in fertility investigations and treatments; and checking on whether or not genes are being expressed in living cells, tissues, and organs. In the food industry, the reaction can be used to test for microorganisms in food. In addition, the luciferin–luciferase reaction is used as a sensor for environmental pollutants.

COMPARE the external anatomy of weevils with that of another insect such as an *ANT, DRAGONFLY, HAWKMOTH, HONEYBEE, HOUSEFLY,* or *LOUSE.* Insects are a highly diverse group of animals, with a wide range of body and wing shapes and colors.

bent near the middle, giving them the appearance of having elbows. The antennae are often called "feelers," though they are in fact organs of smell. In some beetles, particularly those that feed in a restricted space, such as a bored hole, there is a groove called the antennal scrobe on the head at the base of each antenna, allowing the antenna to be folded back over the head and thorax.

A weevil's thorax is divided into three segments: the prothorax, mesothorax, and metathorax. The prothorax is large and is usually able to move freely. The mesothorax and metathorax are fused. The mesothorax is small and bears the elytra. The metathorax is

The weevil's snout

Many beetles have a "snout" projecting forward from the front of the head. Biologists call this the rostrum, and it is a characteristic of the weevils. The snout is formed from two of the plates, or sclerites, on the head (the frons on the front and the vertex on the top) and two additional plates jutting forward underneath the head (the pregula and pregena). All the mouthparts are situated at the end of the snout. Not all insects with a snout are weevils. Some plant bugs have what appears to be a snout, but in fact it is an extension of the mouthparts and not of the head. The weevil's antennae, which are geniculate (bent like a knee or elbow), are situated on the rostrum. At the back of the snout, there is a groove called the antennal scrobe into which the scape (base of the antenna) fits if it is in danger of being damaged.

▶ SNOUT (VIEW FROM BEHIND)
The snout is part of the weevil's head. The mouthparts (maxillae and mandibles) are located at the snout's end. The elbowed antennae (made up of the scape, pedicel, and flagellum) arise partway along the snout.

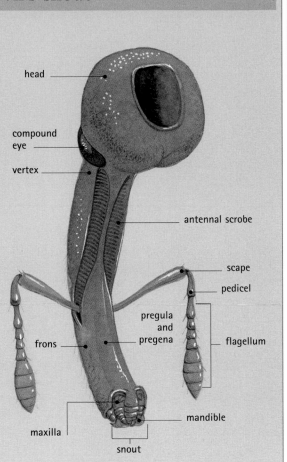

▼ **Jamaican click beetle**
This beetle exhibits bioluminescence—the emission of light by a living organism. Adult males and females use bioluminescence to attract one another at night.

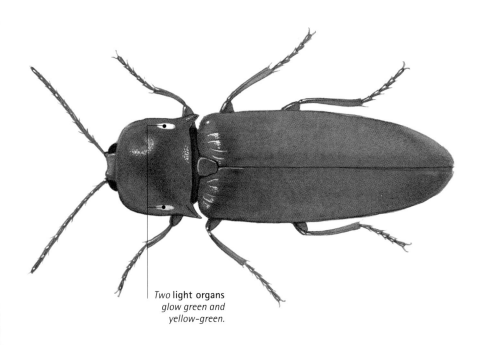

Two **light organs** *glow green and yellow-green.*

large enough to contain the flight muscles and bears the membranous hind wings, which are used for flight.

All three thoracic segments bear a pair of walking legs. These are of standard insect form, the tarsus (lower section of the leg) having five segments, although some families of beetles have fewer tarsal segments. The number of segments is a useful feature for identifying beetles. The abdomen has 11 segments, but in the beetles different segments can be difficult to distinguish. The reproductive organs are situated at the end of the abdomen.

Weevil diversity

Weevils form a very variable group of beetles, and two North American species illustrate this diversity very well. The largest species that lives in the United States is the palmetto weevil, which is a pest of palm trees. Adult palmettos grow to a maximum of 1.2 inches (3.0 cm) from the tip of the snout to the end of the abdomen. Some individuals are all black, and others are red with variable black patterning. The male's snout is covered with tiny bumps, whereas that of the female is smooth and shiny.

Another American species, the sweet potato weevil, is very much smaller, with adults growing to just 0.3 inch (0.8 cm) long. The snout is curved and as long as the thorax. The head is black, the antennae, thorax, and legs are orange to reddish brown, and the abdomen and elytra are metallic blue. Both palmetto and sweet potato weevils are capable of flight.

▶ *The bearded weevil lives in the rain forests of South America. It has very long front legs, and the male has a beard of golden hairs on its snout. The hairs are used to make females receptive to mating.*

◀ *The weevil* Cholus cinctus *lives in the rain forests of Costa Rica. Here, it is feeding on a heliconia plant. The weevil's "feet" give it a good grip on the plant.*

1351

Internal anatomy

CONNECTIONS

COMPARE the main flight muscles of a flying weevil with those of a flying bird such as an *EAGLE*. In birds, the depressor and levator muscles attach to the wing bones. In contrast, the main flight muscles of weevils—the dorsal longitudinal and dorsal ventral muscles—are directly attached not to the wings but to the inside of the thorax.

Internally, weevils are similar to other insects and arthropods: they have muscles and nervous, circulatory, respiratory, digestive, excretory, and reproductive systems. Beneath a beetle's exoskeleton are groups of muscles. Inward-jutting sections of the exoskeleton, called apodemes, serve as attachment sites for muscles. The muscles that operate the insect's jointed legs enable it to walk. Insects usually walk in a fashion called alternating tripod sequence, in which balance is maintained by always having three legs touching the ground.

Muscles for flight

Other muscles in the thorax power the wings during flight. The flight muscles are particularly powerful and require a lot of energy during flight. In most flying insects, including flying beetles, the principal flight muscles are not attached to the wings—they are called indirect muscles. Instead, they are attached to the flexible exoskeleton of the thorax, to which the wings are also attached. One set of muscles—the dorsal longitudinal muscles—runs between apodemes in the winged segment of the thorax and arch the top of the thorax when they contract. This forces the wings down by lever action. The other set of muscles—the dorsoventral—runs from top to bottom. They are antagonistic to the dorsal longitudinal muscles: during contraction, they compress the thorax by pulling down its top, forcing the wings upward. Flying insects also have smaller, direct muscles attached to the base of the wings that adjust their angle during flight. The wing hinge contains an elastic protein called resilin, which allows for fast, repetitive movement. In certain female weevils some of the wing muscles degenerate, and the nutrients from their breakdown are used for the production of eggs.

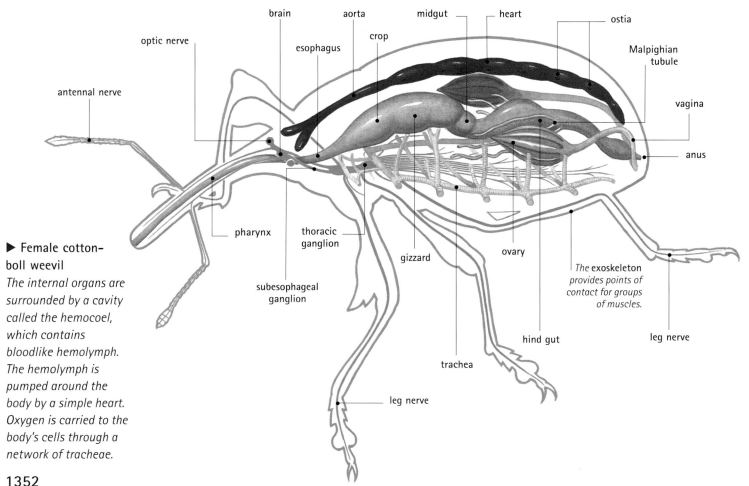

▶ **Female cotton-boll weevil**
The internal organs are surrounded by a cavity called the hemocoel, which contains bloodlike hemolymph. The hemolymph is pumped around the body by a simple heart. Oxygen is carried to the body's cells through a network of tracheae.

brain   aorta   midgut   heart   ostia
optic nerve   crop   Malpighian tubule
esophagus
antennal nerve   vagina
anus
pharynx   thoracic ganglion   ovary
subesophageal ganglion   gizzard   *The exoskeleton provides points of contact for groups of muscles.*
hind gut   leg nerve
trachea
leg nerve

Sensory processing

As in other animals, weevils have a nervous system to receive, transmit, and interpret sensory information from both outside and inside the body and then make an appropriate response such as walking, flying, or releasing a pheromone (chemical signal). The central nervous system of a weevil is made up of a paired nerve cord running along the ventral surface of its body and several attached ganglia (collections of nerve cells, or neurons). Several fused ganglia in the weevil's head form a simple brain.

Hemolymph

The internal organs are surrounded by a cavity called the hemocoel, which contains a bloodlike fluid called hemolymph. This fluid bathes the internal organs, providing nutrients for tissues and cells; and it removes the gas carbon dioxide, which is a waste product of cellular respiration. The hemolymph also removes bacteria and dead cells from body tissues. A simple, tubelike heart, lying close to the dorsal surface of a weevil, helps move the hemolymph around the body. The heart has openings called ostia, and there are also vessels running from the heart. In winged weevils there are also veins in the wings through which hemolymph circulates.

As in other insects, air enters weevils through porelike openings in the exoskeleton called spiracles, which are located on either side of the thorax and abdomen. From there, the oxygen-rich air passes along tubes called tracheae and into finer tubes called tracheoles, which can supply every cell in a weevil's body with oxygen for cellular respiration.

Digestive system

The first part of the digestive system is the foregut, which begins with the mouth. Ingested food then passes into the muscular pharynx, which is the widened beginning of the esophagus. In the metathorax (the third and final section of the thorax), the esophagus expands into the crop, which acts as a food store. The crop opens into the gizzard, which is a small chamber lined by hard ridges and folds, or "teeth." There, food may be ground up by species that chew solid food. In species that take liquid food, the gizzard acts as a filter. The gizzard lies just past the junction between the metathorax and the abdomen. Food passes

through a valve and into the midgut, in which further digestion and absorption take place. The wall of the midgut secretes digestive enzymes onto the food and is protected from damage by a thin membrane. Where the midgut passes into the hindgut, there are four to six Malpighian tubules. These project into the hemocoel and absorb the wastes from bodily functions and those carried by the hemolymph, passing the wastes into the hindgut. The end of the hindgut is the rectum, and there the undigested remains are stored until they are passed from the body through the anus. In some beetles, the hindgut is also important for water absorption.

The foregut and hindgut are lined with cuticle that is shed during molting. Absorbed food is stored in tissues called the fat body, which are comparable to a mammal's liver.

Reproductive organs

Weevils are dioecious—there are two sexes: male and female. The reproductive organs—testes and a penis or aedeagus (in males), and ovaries and a vagina (in females) are located in the rear of the abdomen. The male fertilizes the female internally, and the eggs are fertilized during oviposition (egg-laying).

▲ *The pine weevil lives on conifer trees. Adults feed on shoots, and female adults drill holes in the wood to lay eggs. Pine weevil larvae cause damage to the trees by eating the inner bark.*

Nervous system

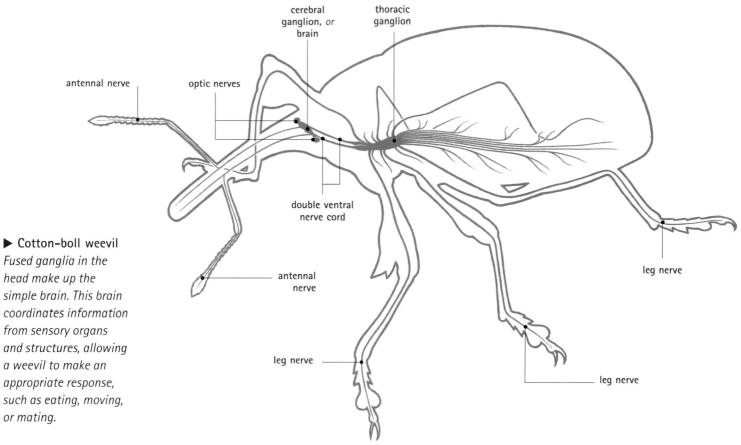

cerebral
ganglion, *or*
brain

thoracic
ganglion

antennal nerve

optic nerves

double ventral
nerve cord

antennal
nerve

leg nerve

leg nerve

leg nerve

▶ **Cotton-boll weevil**
*Fused ganglia in the
head make up the
simple brain. This brain
coordinates information
from sensory organs
and structures, allowing
a weevil to make an
appropriate response,
such as eating, moving,
or mating.*

The central nervous system has a double ventral nerve cord, with ganglia in each body segment that connect the two nerve cords. Each longitudinal nerve cord is called a connective, and the transverse nerves are called commissures. In a weevil's head, several segmental ganglia are fused to form the cerebral ganglion, or simple brain. There are three ganglia in the thorax that tend to be large because of the requirements for flight. In some groups of beetles, the ganglia located in the mesothorax and metathorax may be fused.

Typically, the abdomen shows six ganglia, but these may appear to be reduced in number because adjacent ganglia are fused. In weevils, there are two separate centers of fusion in the abdomen, and in many other beetles such as scarabs and chafers there is just one. Peripheral nerves from each segmental ganglion supply structures within each body segment. These nerves may carry signals in both directions, from sensors to the brain, and from the brain to an effector organ, such as a muscle.

IN FOCUS

The cucujo beetle

The cucujo beetle is a tropical American insect that shows flashes of bioluminescence, which are controlled by nerve impulses. The beetle is a strong flier and lives among tall shrubs and trees in which it darts around, usually in straight lines. It has two types of light organs on different parts of the body, each type producing its own hue: a yellowish green light and a reddish light. The different colors are produced at different times, and the light tends to flash at irregular intervals. This beetle is believed to produce the brightest light in the animal kingdom. Although it produces only about 0.0255 candela (0.025 candle power), the apparent brightness is mainly due to the range of wavelengths it emits—that to which the human eye is most sensitive. The cucujo beetle's steady light is not uniform, but varies at a rate of about five cycles per second and by an amount of 5 to 6 percent of the maximum brightness. This is most likely due to nerve impulses to the light organs that switch on and off. When the impulse is switched off, the light starts to dim. When the next impulse arrives, the light gets bright again.

Circulatory and respiratory systems

As in other beetles, a weevil's circulatory system is mainly an open one, with a single closed dorsal vessel passing from the back end forward and into the head. This dorsal vessel is closed at the rear end and open in the head and has a muscular wall. The rear part (in the abdomen and some or all of the thorax) constitutes the heart, which is divided into several segments with attached muscles. In each segment there is a pair of openings, called ostia, through which the bloodlike hemolymph enters the heart. The ostia also act as valves, allowing one-way movement of hemolymph through the heart. A wave of contraction of the muscular wall of the heart starts in the rear and passes forward. Other ostia allow the hemolymph to leave the heart and move into the body segments during this wave of contraction. The front part of the dorsal vessel is the aorta, which divides into two branches in the head.

Hemolymph in the heart is pumped forward through the aorta and into the head. At the junction of the heart and aorta are the aortic valves. These valves prevent the hemolymph from flowing backward. In this way, it is pumped around the body, but not in a series of vessels. Instead, the hemolymph surrounds the body organs, filling the body space, or hemocoel. Hemolymph supplies water and food molecules to the organs and takes away waste products, including carbon dioxide from cellular respiration. Hemolymph plays little part in the transport of oxygen.

Respiratory system

The breathing, or respiratory, system consists of a system of tubes that pass throughout the body. On the sides of most segments of the abdomen and thorax are openings called spiracles. The weevil can control the size of these openings. The spiracles open into a pair

CONNECTIONS

COMPARE the open circulation of a weevil with the closed circulation of a different type of invertebrate, an OCTOPUS. In beetles, hemolymph bathes the internal organs. However, in octopuses there is a closed system of blood vessels that supply all the tissues with oxygen and nutrients.

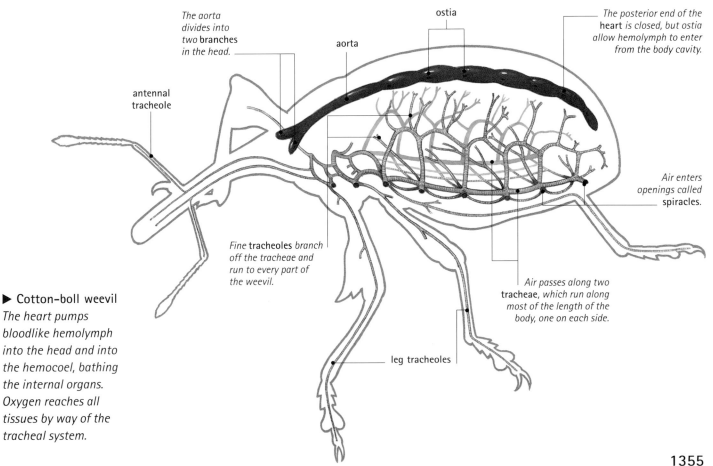

The aorta divides into two branches in the head.

ostia

aorta

The posterior end of the heart is closed, but ostia allow hemolymph to enter from the body cavity.

antennal tracheole

Air enters openings called spiracles.

Fine tracheoles branch off the tracheae and run to every part of the weevil.

Air passes along two tracheae, which run along most of the length of the body, one on each side.

leg tracheoles

▶ Cotton-boll weevil
The heart pumps bloodlike hemolymph into the head and into the hemocoel, bathing the internal organs. Oxygen reaches all tissues by way of the tracheal system.

of main tubes, one on each side, that run front to back for most of the length of the body. These two tubes have elastic walls and are called the tracheae. From these, finer and finer tubes, or tracheoles, branch off and end at individual cells, especially muscle cells and other cells that have a high need for energy. The walls of the tracheal system are lined with cuticle. This lining is called the intima and is shed when the weevil molts. Air containing oxygen passes along the tracheae and tracheoles to the cells, where it is used in respiration. In more active species the air may be pumped through the system by synchronized muscular contractions of the body. The movement of oxygen close to the cells is by passive diffusion.

▲ *A ladybug's membranous hind wings are used for flight and have a much larger surface area than the patterned wing cases (elytra). The hind wings have veins through which hemolymph flows. Ladybugs are members of the family Coccinellidae and are closely related to weevils.*

Carbon dioxide removal

Carbon dioxide can be removed through the tracheal system by the process in reverse, but most carbon dioxide is lost in solution through the hemolymph and excretory Malpighian tubules. One of the problems associated with the uptake of oxygen is that water loss also occurs, and the tracheal system is the major source of water loss from weevils and other beetles.

Reproductive system

There are two general types of reproductive systems in male beetles, depending on the structure of the testes. In beetles in the suborder Adephaga (adephagan beetles), the testes are a simple tubular structure that becomes the vas deferens. In other beetles, including weevils, each testis is a number of separate follicles (small sacs or chambers) clustered around a common duct that becomes the vas deferens (plural, vasa deferentia). The vasa deferentia (one from each testis) meet to form a common ejaculatory duct, and this passes to the outside through the penis, or aedeagus.

Within a female beetle's reproductive system there are two ovaries, each containing a number of ovarioles (the tubes in which the eggs develop). The female system also shows two generalized structural types, depending upon the nature of the ovarioles in the ovaries. In the polytrophic type, as in adephagan beetles, the developing egg cells are in a line in the ovariole. The nutritive cells, which help the egg cells develop, are situated in groups between the egg cells. The acrotrophic type is seen in other beetles, including weevils. In this type, the nutritive cells lie at the ends of the ovarioles, and the egg cells are below them. The ovarioles pass the fully formed eggs into the two oviducts, one from each ovary. The eggs may be retained there before being fertilized and passed out through the vagina. The female also has a sac, called the spermatheca, in which sperm are stored after mating. These sperm are used to fertilize the eggs as they are laid. Beetles generally mate and use internal fertilization.

Males are usually smaller and narrower than females, which also have a more rounded abdomen and therefore more rounded elytra. Other sexual differences exist, such as the longer snout of the female nut weevils and the elaborate horns of male stag beetles. When mature and ready for mating, the female beetles produce pheromones, chemicals that pass into the air and attract the males. There are other attracting signals, such as sound or light. Once the sexes are together, the male will try

CONNECTIONS

COMPARE fertilization in weevils with fertilization in an *EARTHWORM, COELACANTH, HAMMERHEAD SHARK, TORTOISE, OSTRICH, RAT,* and *ELEPHANT.* All these animals practice internal fertilization, which maximizes the chance of fusion of sperm and eggs.

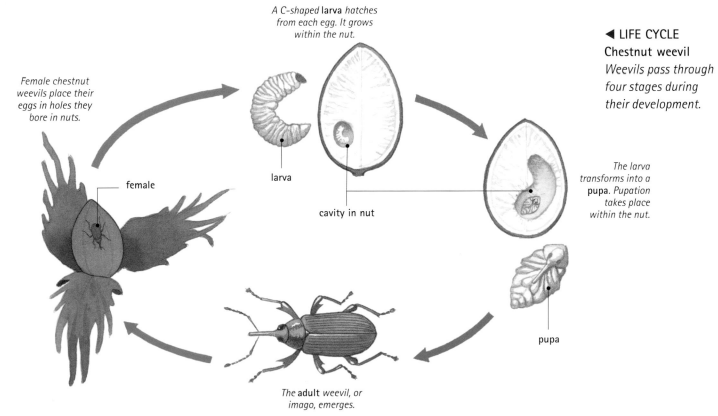

A C-shaped **larva** *hatches from each egg. It grows within the nut.*

Female chestnut weevils place their eggs in holes they bore in nuts.

female

larva

cavity in nut

◄ LIFE CYCLE
Chestnut weevil
Weevils pass through four stages during their development.

The larva transforms into a **pupa.** *Pupation takes place within the nut.*

pupa

The **adult** *weevil, or imago, emerges.*

1357

◀ *Weevils of the genus* Eupholis, *from Papua New Guinea, mate on a leaf. In beetles, like almost all terrestrial animals, the female's eggs are fertilized internally.*

IN FOCUS

Life cycle

Weevils and other beetles have a life cycle that shows complete metamorphosis: they have a pupal phase between the larva and the adult. Weevils pass through four stages during their development. The eggs are usually oval and rarely show any special features. The larvae that hatch from the eggs have a well-developed head with biting mouthparts. They show a range of body forms, depending on their lifestyle. The basic type has legs and is active. Modification tends to be along two lines: the reduction and loss of legs and a curving of the body. The larvae of weevils that lay eggs on food plants do not have legs (they are apodous) and are curved like a letter C. This body structure is very obvious in species that feed within the roots and fruits of plants. In such environments, the larvae do not need legs, since they excavate their way through the plant and do not need to walk to find food. In general, the larvae pass through about four stages, molting between each, before pupating. Pupation often takes place in the soil. The mature larva usually prepares a cell in which to pupate. The pupal phase may last from a few days to a year or more. At the end of this time the adult, or imago, emerges.

to climb onto her back. Then, he tries to stimulate her so mating can take place. This stimulation takes a variety of forms, such as movements of the antennae or legs or the production of other pheromones. When the female is receptive, the male inserts his aedeagus into her vagina and ejects sperm. The pair may remain in this copulatory state for a number of hours. Once mating is over, the male seeks another partner.

Nut weevils and egg-laying

In nut weevils the characteristic feature of weevils—the snout—is very long and slender. This shape is particularly obvious in the female, in which the downward-curving snout can be nearly twice the length of the body. In the male nut weevil the snout is much shorter. The antennae are situated at a distance from the base of the snout, equivalent to about a third to half of the body length.

The female's snout needs to be long, slender, and also flexible because of the way in which she lays her eggs. The mandibles are at the end of the snout, and many weevils use them to cut a deep hole into a food plant for laying eggs. She then turns around and lays an egg in the hole. The ovipositor, or egg-laying tube, is about the same length as the snout but is contained within the abdomen. The ovipositor is extended to lay the egg.

Nut weevils have a long snout that can drill through thick nut walls. The weevils drill at an angle to get to the softer tissue in the developing nut. It is there that the egg is laid and the larva starts to feed. Usually, the larva is off-white with a darker head and no legs. Also, it is slightly curved. A female nut weevil will lay about 40 eggs, each one in a separate nut, so she has to drill many holes—a time-

▲ Black vine weevil larvae have a brown head, a fat translucent body, and no legs. They feed on roots and storage organs, such as begonia corms. They often overwinter as larvae.

consuming task. As the female drills deeper into the nut, the first segment of the antenna, the scape, is pushed into a backward-directed groove called the antennal scrobe on the snout. This arrangement keeps the antennae out of the way and safe from damage.

The thicker the nut wall, the earlier in the nut's development the female lays the eggs. The female hazelnut weevil lays her eggs in the late spring, whereas the acorn weevil lays hers later, in the summer. Infected nuts fall off the tree sooner than the noninfected ones.

When mature, a larva bites its way out of the nut, leaving a hole, and burrows into the ground. There, the larva makes a spherical cavity in which it overwinters. It usually pupates the following spring but may stay in the cell for more than one winter. Later, the adult emerges from its pupa and exits the cell to feed and mate, or it hibernates again, emerging the following spring.

NICK HOLFORD

Snout drills hole.

ovipositor

▲ Chestnut weevils
Female chestnut weevils use their snout to drill holes in a developing nut. They then insert their ovipositor into the holes, leaving one egg in each hole.

FURTHER READING AND RESEARCH
Jackson, T., and J. Martin (eds.). 2003. *Insects and Spiders of the World*. Marshall Cavendish: Tarrytown, NY.
A Field Guide to Boll Weevil Identification: msucares.com/pubs/techbulletins/tb0228.pdf

Wildebeest

CLASS: Mammalia ORDER: Artiodactyla
SUBORDER: Ruminantia FAMILY: Bovidae
GENUS: *Connochaetes*

The two species of wildebeests are among the most conspicuous animals of the African savanna. Both species form large, often nomadic herds that roam the plains in search of fresh grass, often in the company of zebras. The common, or blue, wildebeest, also called the brindled gnu, is a large, highly social antelope. It is the most abundant large grazing mammal on the great savanna plains of East Africa. The black wildebeest, which lives only in South Africa, is a little smaller than its close relative.

Anatomy and taxonomy

Scientists categorize all organisms into groups, or taxa, based on anatomical, biochemical, and genetic similarities and differences. The family Bovidae contains 140 species of large, hoofed herbivores, including cattle, gazelles, goats, sheep, and the two species of wildebeests.

● **Animals** Animals, including common and black wildebeests and their relatives, are multicellular organisms that consume organic material from other organisms to fuel their metabolism. Animals differ from other multicellular life-forms in their ability to move around (mostly by using muscles) and their rapid reaction to stimuli such as touch, light, and certain chemicals.

● **Chordates** At some stage in its life cycle, a chordate has a stiff, dorsal (back) supporting rod called the notochord that runs all or most of the length of the body.

▼ *The two species of wildebeests are the only members of the genus* Connochaetes, *which is part of the diverse family Bovidae: ruminant, even-toed ungulates that have horns. Close relatives include other antelope such as the impala and hartebeest.*

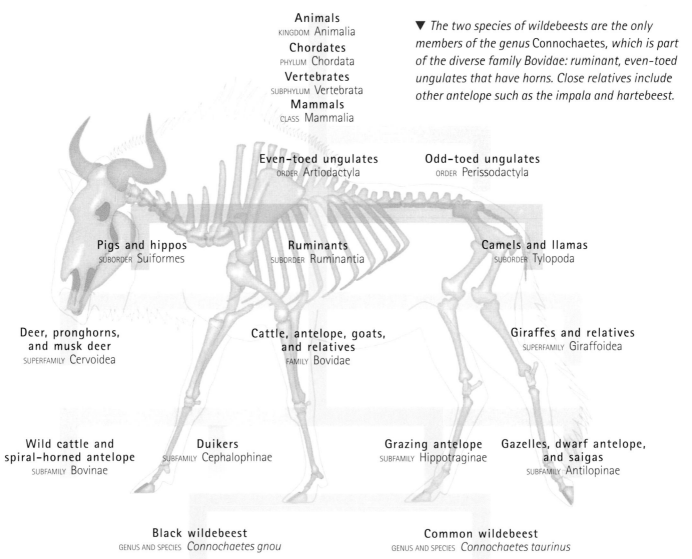

Animals
KINGDOM Animalia

Chordates
PHYLUM Chordata

Vertebrates
SUBPHYLUM Vertebrata

Mammals
CLASS Mammalia

Even-toed ungulates
ORDER Artiodactyla

Odd-toed ungulates
ORDER Perissodactyla

Pigs and hippos
SUBORDER Suiformes

Ruminants
SUBORDER Ruminantia

Camels and llamas
SUBORDER Tylopoda

Deer, pronghorns, and musk deer
SUPERFAMILY Cervoidea

Cattle, antelope, goats, and relatives
FAMILY Bovidae

Giraffes and relatives
SUPERFAMILY Giraffoidea

Wild cattle and spiral-horned antelope
SUBFAMILY Bovinae

Duikers
SUBFAMILY Cephalophinae

Grazing antelope
SUBFAMILY Hippotraginae

Gazelles, dwarf antelope, and saigas
SUBFAMILY Antilopinae

Black wildebeest
GENUS AND SPECIES *Connochaetes gnou*

Common wildebeest
GENUS AND SPECIES *Connochaetes taurinus*

● **Vertebrates** In vertebrates, the notochord develops into a backbone (spine or vertebral column) made up of vertebrae. The muscular system that moves the head, trunk, and limbs of a vertebrate consists primarily of muscles in mirror-image arrangement on either side of the backbone (bilateral symmetry about the skeletal axis).

● **Mammals** Mammals are warm-blooded vertebrates that have hair made of keratin. Females have mammary glands that produce milk to feed their young. The mammalian inner ear contains three small bones (ear ossicles), two of which are derived from the jaw mechanism of the group's reptilian ancestors.

● **Placental mammals** These mammals nourish their unborn young through a placenta, a temporary organ that forms in the mother's uterus during pregnancy.

● **Ungulates** Hoofed mammals include most large grazing and browsing herbivores in the order Perissodactyla (odd-toed ungulates, which includes horses, rhinos, and tapirs); and as well as even-toed ungulates, which includes antelope, cattle, camels, and deer. Hoofed mammals have limbs suited to running, with nails modified into hard hooves. Ungulates have long foot bones and, on all four feet, a reduced number of digits and no big toe.

▲ *A common wildebeest crosses a shallow pool. Both sexes bear horns and have black fur on their face, mane, and tail.*

● **Artiodactyls** Even-toed ungulates have two or four toes on each foot. In all but the hippopotamuses, only the third and fourth digits are weight-bearing. The group includes various types of deer, camels and llamas, pigs, peccaries, giraffes, hippos, and the pronghorn, as well as the antelope and their cousins, the cattle, sheep, and goats.

● **Bovidae** The members of this large family are horned ungulates—in all species, males have horns; often females do, too. Horns are permanent bony outgrowths from the frontal bones of the skull, unlike the antlers of deer, and are encased in a sheath of the horny material keratin. The bovids are divided into several subfamilies, which include the Bovinae (wild cattle and spiral-horned antelope), Cephalophinae (duikers), Hippotraginae (grazing antelope, including the wildebeests), and Antilopinae (gazelles, dwarf antelope, and saigas).

● **Bovinae** This subfamily of the Bovidae includes the cattle and spiral-horned antelope, such as kudus and elands. Common cattle were domesticated up to 10,000 years ago.

● **Cephalophinae** The subfamily is made up of duikers, small to medium-size antelope that live mostly in dense forest thickets. Some are little bigger than a domestic cat.

● **Hippotraginae** The subfamily of grazing antelope contains 23 species in 11 genera, including the wildebeests, reedbucks, oryxes, and impala.

EXTERNAL ANATOMY Wildebeests have a muscular body covered with short blue-gray to brown hair. The head is large and bears impressive curved horns in both sexes. *See pages 1362–1367.*

SKELETAL SYSTEM Wildebeests have a light frame suited to running. The thoracic vertebrae have huge spines to support shoulder muscles. The horns have a bony core growing from the frontal bones of the skull *See pages 1368–1369.*

MUSCULAR SYSTEM The neck and shoulders are massively muscled, forming a large hump. The rump is also muscular. *See pages 1370–1371.*

NERVOUS SYSTEM Wildebeests have a small brain, good eyesight, sharp hearing, and an acute sense of smell. They rarely sleep. *See pages 1372–1373.*

CIRCULATORY AND RESPIRATORY SYSTEMS A large heart and deep lungs allow wildebeests to run for long periods. *See page 1374.*

DIGESTIVE AND EXCRETORY SYSTEMS As ruminant herbivores, wildebeests have a very large, multichamber stomach in which symbiotic bacteria break down tough plant material. *See pages 1375–1377.*

REPRODUCTIVE SYSTEM Adult males and females look similar. Females give birth to a single, precocious calf and suckle it on rich milk. *See pages 1378–1381.*

External anatomy

COMPARE the horns of wildebeests with the antlers of the **RED DEER**. Wildebeests' horns are permanent features, but the red deer's antlers are lost and regrown each year.

COMPARE the beard and mane of wildebeests with those of the **LION**, **WOLF**, and **ZEBRA**. In all species, the extra hair makes the animal look larger and more powerful. Wildebeests toss their mane during displays aimed at intimidating rivals.

The common, or blue, wildebeest is one of the largest species of antelope, weighing 330 to 640 pounds (200–290 kg). Unlike most other antelope, wildebeests are heavyset and look clumsy. The legs are slender, but the body is bulky, with a deep chest, huge shoulder hump, and downward-sloping back. The neck, chest, and abdomen are deep, and the head is large.

The body is covered with a coat of short, glossy hair that varies considerably in color, from blue-gray to dark brown or pale fawn. The legs are usually brown or a dirty beige. The tail bears a long tuft of dark hairs. Whatever the fur color, the face and muzzle are always black. A mane of long black hair runs along the crest of the neck from the top of the head to the

▶ **Common wildebeest**
An adult male common, or blue, wildebeest is one of the largest antelope. Oxlike in appearance, wildebeests have heavy, curved horns on a massive head with a broad muzzle. The forelegs are heavier than the hind legs.

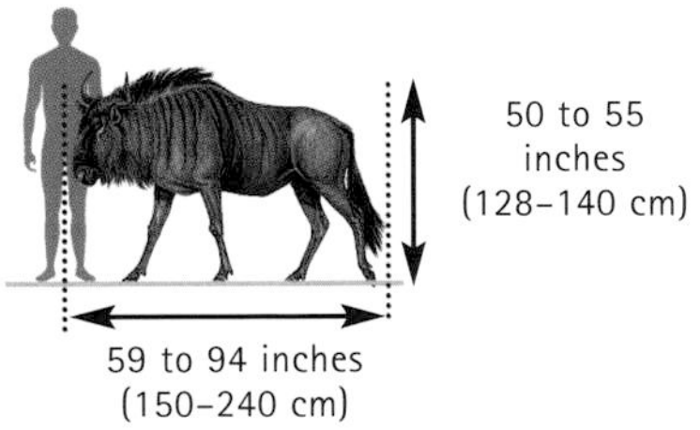

The **ears** *are large and have acute hearing to detect predators such as lions. The ears turn downward during courtship.*

The **mane** *is long, tufted, and black. Wildebeests often shake their mane during ritual displays.*

The **head** *is massive with a broad muzzle. Scent glands on the face are rubbed on the rump of other wildebeests during courtship.*

The **legs** *are long and elegant. They allow the wildebeest to run fast, especially when being chased by a predator.*

The **hooves** *are cloven with two slightly splayed toes, an arrangement that helps stabilize the weight of this large animal.*

50 to 55 inches (128–140 cm)

59 to 94 inches (150–240 cm)

middle of the shoulder hump. The mane is not erect like that of a zebra but is soft and flows like that of a domestic horse. A beard of long, pale hairs grows from the chin, throat, and neck and sometimes extends right down to the chest. Black or dark brown stripes on the neck and flanks enhance the overall effect of the flowing mane and beard.

The large head looks a little like that of an ox, with a long, broad face and a slightly convex (outwardly curved) muzzle. The eyes are large, slightly bulging, and placed wide apart on the sides of the head. The ears are long and

*The **body** appears front-heavy. There are black or dark brown stripes on the neck and the flanks.*

*The **tail** is black and conspicuous. The animal often cavorts, lashing its tail from side to side. This behavior may help attract a mate.*

*Wildebeests are digitigrade—they walk and run on their **toes**.*

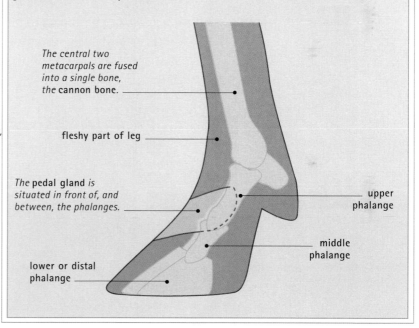

▶ **Eland**
There are two species of elands: the common eland and the giant eland, both of which are in the genus Taurotragus. *Along with species of antelope in the genus* Tragelaphus, *such as the kudu and bongo, elands are members of a group called spiral-horned antelope.*

located on the sides of the head, just below the horns. The ears typically droop slightly, giving the animal a relaxed expression. In fact, wildebeests are highly alert.

Impressive horns

By far the most impressive features of the head are the horns. In all species of the bovid family, the males bear horns, and in more than half of all species, so do the females. The horns of bovids vary considerably in shape and size, but all have a similar internal structure. Horns have a bony core with a tough outer sheath of a horny protein material called keratin, which is secreted by the epidermis. The bony core of bovid horns develops from the frontal bones of the skull. In most species, the core continues to grow throughout life, although very slowly in

▼ *A vast herd of wildebeests migrates, shown below crossing the dried-up Mara River in Kenya. They migrate to find prairies with fresh, long grass.*

Horn shape

Bovid horns vary greatly in size and shape. The differences reflect behavior or habitat. Forest-dwelling bovids usually have small horns; if they are long they tend to curve backward to avoid snagging on branches. When long-horned species run through trees they have to tilt the head back to keep their horns clear of branches. The spiral horns of the kudu are used for ritualized interactions. The horns can also be used as weapons: for butting rivals as in male bighorn sheep and sable or even goring other animals as in water buffalo. However, real violence rarely happens without provocation and usually takes place in self-defense or in the defense of young animals. The four-horned antelope is unique among bovids in having four horns instead of two.

▼ Sable
The long backward-curving horns of the sable are used as means of self-defense against predators, such as lions. Both sexes grow horns, but they are longer in males, reaching 65 inches (1.7 m).

▶ Kudu
Male kudus use their horns for wrestling to establish dominance hierarchies. The horns of two wrestling males can interlock permanently, eventually resulting in the death of both males.

◀ Four-horned antelope
The two pairs of horns are smooth and conical. The main pair, just in front of the ears, grows to 2 to 5 inches (5–12 cm) long. The shorter pair may fall off in older animals. Only males have horns.

old animals. The horns are never branched, and unlike the antlers of deer they cannot be cast off or grown back if damaged.

The horns of wildebeests grow up to 31 inches (80 cm) long and are thick and heavy, especially in males. The horns of females are slightly flat in cross section, whereas those of a mature male are round in cross section. The horns grow from separate knobby bases called bosses, which are larger in males than in females. Each horn is strongly curved—angled first out to the side of the head and then curving upward and slightly inward again. Like the horns of cattle and oxen, wildebeest horns have a smooth surface, with no rings or spiral ridges.

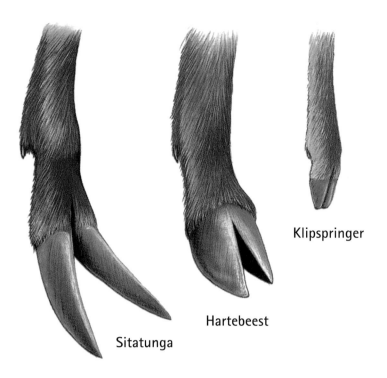

Sitatunga

Hartebeest

Klipspringer

Antelope have two toes, each with a hoof, giving a cloven appearance. The sitatunga and hartebeest are relatively large antelope with splayed hooves that support their weight on soft ground. The mountain-dwelling klipspringer, a dwarf antelope, has small, pointed hooves for balance.

and walk on the tips of their toes. The ungulates evolved from ancestors that had five digits on each foot, but in every living species of ungulates this number has been reduced. In the odd-toed ungulates (order Perissodactyla) the number has been reduced to three toes per foot as in rhinoceroses and tapirs or just one as in horses. The even-toed ungulates have either two or four toes.

The chief benefit of walking on the tips of a reduced number of toes is speed. Ungulates are able to run fast. Animals run fast for two reasons: because they are hunters or because they are hunted. Like all ungulates, wildebeests are herbivores. Their food does not need to be chased, but both species of wildebeests are often targeted by predators such as lions, hyenas, leopards, jackals, wild dogs, and crocodiles. Even birds of prey sometimes make off with very young calves.

Both species of wildebeests have two weight-bearing toes on each foot, each toe with a small, hard hoof. Like human fingernails, hooves are

Walking on tiptoes

Along with other antelope, cattle, camels, and deer, wildebeests are even-toed ungulates. The word *ungulate* comes from the Latin word *unguis*, meaning "nail," "hoof," or "claw." All ungulates have toenails that have evolved into hooves; and, unlike humans or bears, which walk on the soles of the feet, ungulates stand

Three species of reedbucks live in savanna, mountains, forest, and reed beds in Africa. They bear short, curved, ridged horns.

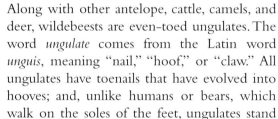

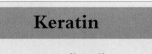

IN FOCUS

Keratin

Keratin is an extraordinarily versatile protein. Not only is it the main constituent of horn; it is also the vital compound in all mammalian hair, bird feathers, and reptilian scales. It forms the hooves, nails, and claws of all kinds of different vertebrates. Keratin is also produced in the skin in response to sustained pressure or abrasion, leading to the formation of horny calluses. People develop keratinized calluses on the soles of the feet and sometimes on the hands if they perform long hours of manual labor. Wildebeests develop calluses on the legs and the brisket—the part of the breast that makes contact with the ground when the animal rests.

▶ **Male impala**
Only males have S-shape horns, which are heavily ridged, are relatively thin, and have ends that are far apart. Impalas live in Africa.

made of keratin, or horn. The hooves grow slowly but continually, compensating for gradual wear and tear.

In addition to the true hooves, each foot has two false hooves, one on each side. These are the vestiges (degenerated body parts) of the second and fifth toes. The false hooves do not make contact with the ground.

Different species of antelope have dissimilar shaped toes. Those of small, athletic species, such as the impalas and the klipspringer, are closely united, and the hooves are small and pointed. These allow the animals to leap and turn very rapidly while maintaining balance. In larger species, such as wildebeests and the

hartebeest, the toes are slightly splayed to spread the animal's weight over a larger area. This feature is taken to an extreme in the swamp-dwelling sitatunga, whose hooves are long and spread very widely, allowing it to walk on soft, boggy ground without sinking.

Wildebeests are not graceful animals. They run with a peculiar rocking-horse gait, not unlike that of giraffes. To lie down, a wildebeest drops first to its front knees, then folds its hind legs to one side as it lowers its rump. Calves sometimes sleep flat out on their side, with legs stretched out straight, but such a relaxed posture is unusual in healthy adults, which rest upright on the sternum, or breastbone.

▲ **Female four-horned antelope**
Bearing four horns is considered a "primitive" feature in antelope. The four-horned antelope, also called the chousingha, is found in India.

◀ *The curved, oxlike horns can be clearly seen in this group of wildebeests drinking from a river in the Serengeti, Tanzania.*

Skeletal system

COMPARE the limb bones of a wildebeest with those of a *CHIMPANZEE*. The chimp has many more bones in the arms, legs, hands, and feet, which allow it to make a much greater range of movements than the wildebeest. However, a chimpanzee is not nearly such a good runner as a wildebeest.

Wildebeests are front-heavy animals, and this structure is very much reflected in the skeleton. The skull is very large and, with the addition of massive horns, very heavy. Because of this weight, the neck has to be very strong. The massive neck and shoulder muscles require large bones for anchorage. The cervical, or neck, vertebrae are therefore large, with obvious knobby processes (extensions) on all but the first two (the atlas and axis bones, which allow for movement of the head). More noticeable still are the long dorsal processes, or spines, that protrude from the thoracic, or chest, vertebrae. These processes form a rounded crest above the shoulders and serve as attachment points for the muscular hump.

The ribs of bovids are typically broad, and the gaps between them are smaller than in many other mammals. This arrangement makes the rib cage strong, so it is not distorted too much by the action of fast running.

Compared with the forequarters, the pelvic girdle is much more lightly built. The limbs, too, are surprisingly gracile (delicate) for such a heavy animal. However, the upper leg bones (the humerus and femur) are embedded within the soft tissues of the trunk, giving the animal a stocky appearance.

The limb bones provide strength and speed, but at the expense of flexibility. For example, the ulna and fibula are small, so the animal takes all its weight through the large radius and

▼ Common wildebeest
The skeleton of a wildebeest is built for stamina and speed. The skull is heavy, the pronounced dorsal processes of the thoracic vertebrae attach to large muscles, and the limbs are long for fast running.

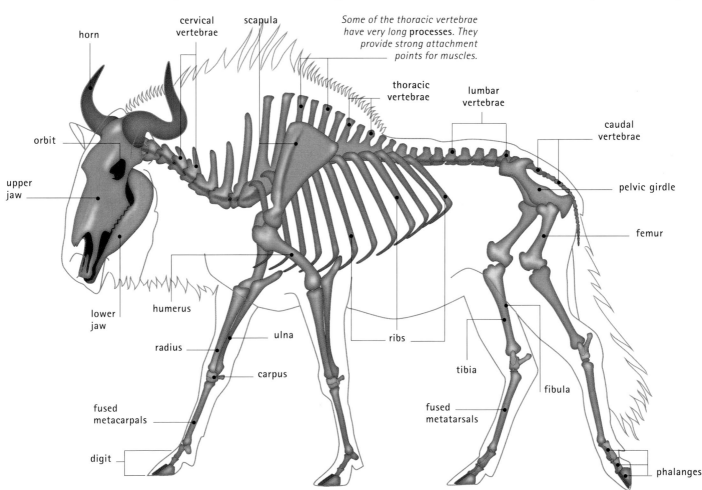

Some of the thoracic vertebrae have very long **processes**. *They provide strong attachment points for muscles.*

horn
cervical vertebrae
scapula
thoracic vertebrae
lumbar vertebrae
caudal vertebrae
orbit
pelvic girdle
upper jaw
femur
lower jaw
humerus
ulna
ribs
tibia
radius
fibula
carpus
fused metatarsals
fused metacarpals
digit
phalanges

Fused feet

The most significant characteristic of the ungulate skeleton is the feet. The long, small bones of the feet (the metatarsals and metacarpals) are fused into a single bone—the cannon bone. The foot is carried off the ground, and the animal bears its weight on just two functional digits on each foot.

▼ ANTELOPE FORELIMBS

An antelope's forelimbs evolved over millions of years and provide the animal with speed and endurance. The two outer metatarsals have been lost, and the central two have fused into the long cannon bone.

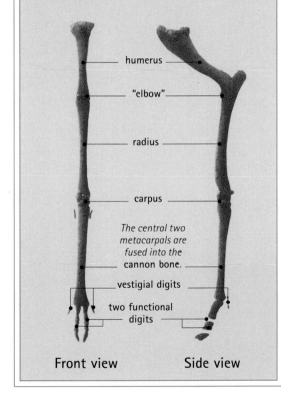

humerus

"elbow"

radius

carpus

The central two metacarpals are fused into the cannon bone.

vestigial digits

two functional digits

Front view   Side view

At the front of the lower jaw, there is an even arc of eight chisel- or spade-shape teeth: three pairs of incisors flanked by a single pair of similarly shaped canines. Unlike some artiodactyls (mouse deer and some primitive members of the deer family) wildebeests have no upper canines or tusks. There are short rows of cheek teeth on each side: three molars and three premolars in the upper jaw; and three molars and two premolars in the lower. The cheek teeth are large, with convoluted surfaces ideally suited to grinding and mashing plants.

Skulls

All bovids have a long skull with a large nasal cavity (reflecting the importance of smell in most species) and a braincase (cranium) that is small relative to the size of the animal. The horns are permanent but do not begin to develop until the young has been born: even small horns would cause problems during birth. The same is true of most artiodactyls, but not of giraffes—their young are born with small horns that fold at the base to allow the head to pass though the birth canal.

▶ WILDEBEEST SKULL

The skull is long with a large nasal cavity and small braincase. There are no teeth at the front of the upper jaw; there is just a bony ridge covered with horny skin used for biting vegetation.

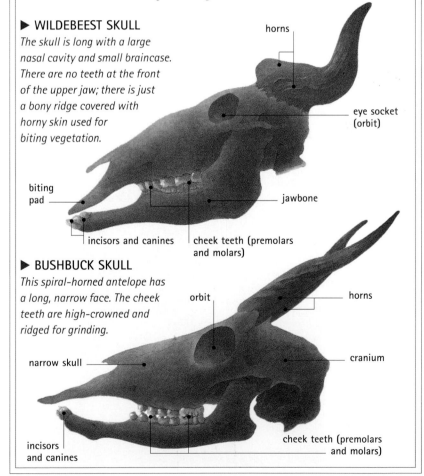

horns

eye socket (orbit)

biting pad

jawbone

incisors and canines

cheek teeth (premolars and molars)

▶ BUSHBUCK SKULL

This spiral-horned antelope has a long, narrow face. The cheek teeth are high-crowned and ridged for grinding.

orbit

horns

narrow skull

cranium

incisors and canines

cheek teeth (premolars and molars)

tibia. Ungulates have one large bone in the lower leg instead of two moderate-size bones (as in primates and carnivores, for example). This arrangement of bones prevents ungulates from rotating their limbs. Thus the limbs can be used only for locomotion, kicking, and scratching.

Incisors, premolars, and molars

Wildebeests lack upper incisors. There are no teeth at the front of the upper jaw; there is just a hard ridge of bone covered with horny skin.

Muscular system

CONNECTIONS

COMPARE the slender legs of a wildebeest with the muscular limbs of an *ELEPHANT*. A wildebeest's legs are streamlined for speed, with elastic tendons and ligaments to put a spring in the step. In contrast, an elephant's pillarlike legs are suited to weight-bearing.

Wildebeests are powerful, muscular animals. In particular, the neck and forequarters are highly muscular, as also in buffalo. Strong muscles support the weight of the large head. The principal muscles of the forequarters are the trapezius, rhomboideus, latissimus dorsi, serratis ventralis, and pectorals. All are relatively large muscles compared with those of most other ungulates, and wildebeests often show off their physique to best effect. Most displays and defensive postures involve lowering the head and turning slightly to one side.

The abdomen is also deeper than in many other ungulates, but most of the bulk is taken up by the digestive system, not by muscle. The flanks and belly are enclosed in a thin tight sheath of oblique and longitudinal muscles. These muscles help keep the vital organs of the digestive and reproductive systems in place.

Internally, the chest and abdomen are separated by a thin sheet of elastic muscle called the diaphragm. The diaphragm is connected by ligaments to the lumbar vertebrae, the ribs, and the breastbone (sternum). The contraction of the diaphragm puts pressure on the lungs, causing the animal to breathe out. When the diaphragm relaxes, the lungs expand and the animal inhales. Therefore, it is important that the diaphragm forms an airtight seal. Thus all the major continual structures that pass from the chest region to the abdomen (the spinal cord, esophagus, and major blood vessels) have to be located above the diaphragm.

▼ Common wildebeest
Wildebeests are muscular, especially in the neck region to support the large head, and in the forequarters to power the forelimbs for running.

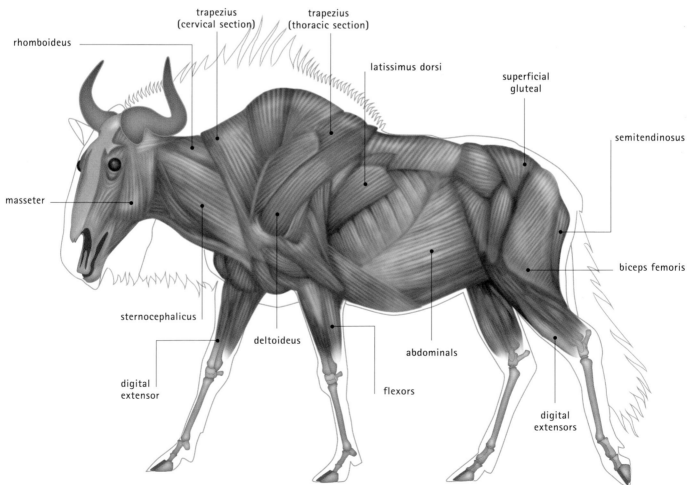

rhomboideus

trapezius (cervical section)

trapezius (thoracic section)

latissimus dorsi

superficial gluteal

semitendinosus

masseter

sternocephalicus

deltoideus

digital extensor

flexors

abdominals

biceps femoris

digital extensors

A spring in the step

Pronking is an unusual kind of movement, highly characteristic of antelope, in which the animal leaps vertically. This leaping is possible only because of the elasticity of the ligaments (bands of connective tissue that hold bones together at joints) in the legs. The ligaments relax when the legs are straight. At the end of a leap, the legs flex as they make contact with the ground, and the ligaments are stretched like rubber bands. The legs then straighten, and the animal is propelled almost effortlessly back into the air.

Slender limbs

All antelope and most ungulates have relatively slender legs. The bulky muscles that power them are restricted to the trunk and connect to the bones by means of long, stretchy tendons. This arrangement is advantageous in several ways. It provides strong leverage, so a relatively small movement of muscle at the top of the leg translates to a wide swing at the bottom. Because the legs are relatively light, less energy is required to swing them. Their slenderness also enhances streamlining.

This complex arrangement of bones is held together by tough, fibrous ligaments—the ligaments around the carpal joint are particularly intricate. The carpal joint is equivalent to the human ankle, but in bovids it is located almost halfway up the leg.

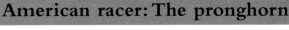

American racer: The pronghorn

Wildebeests have no immediate cousins in the Americas—the true antelope are restricted to the Old World. However, North America is home to a very special relative of the African antelope: a unique animal called the pronghorn. This remarkable ungulate is the only species in the family Antilocapridae. It is named for its unusually shaped horn and famous for its speed. An adult can maintain speeds up to 40 miles per hour (65 km/h) for long periods. The fastest speed ever recorded for a pronghorn over a short distance was an incredible 61 miles per hour (97 km/h).

▼ Pronghorn
This ungulate is the swiftest-running animal in the Americas. Its limbs are long, slender, and very agile, allowing it to cover 27 feet (8 m) in a single leap.

Nervous system

CONNECTIONS

COMPARE the wraparound vision and horizontally long pupils of a wildebeest with the pupils of the *LION*, which close to form circular points. Wildebeests need to be aware of any approaching predators and so require a wide field of vision.

As with all vertebrates, the nervous system of wildebeests and their relatives includes a central nervous system, consisting of the brain and spinal cord; and a peripheral nervous system, the nerves of which branch out all over the body. All peripheral nerves ultimately lead to or from the central nervous system, which processes information from the sense organs and coordinates body functions such as movement. A mammalian body is highly complex. Most of the essential processes of life, such as heartbeat and breathing, take place automatically and are under the control of the part of the nervous system called the autonomic nervous system.

Wildebeests are timid but inquisitive animals. Sometimes, this combination makes them seem almost paralyzed with indecision because the sight of potential danger causes them to stand transfixed on the spot, torn between the urge to get a better look and the urge to get away as fast as possible. The same "fight-or-flight" problem occurs when two rival males perform what scientists studying animal behavior call conflict behavior. For example, each male will often adopt a confrontational stance with the head lowered and turned toward his rival, but with the body leaning away or even making

▼ **Common wildebeest**
Like other mammals, wildebeests have a central nervous system (CNS; brain and spinal cord) and peripheral nervous system (nerves that connect to the CNS). The parts of the brain processing olfaction (sense of smell) are particularly well developed.

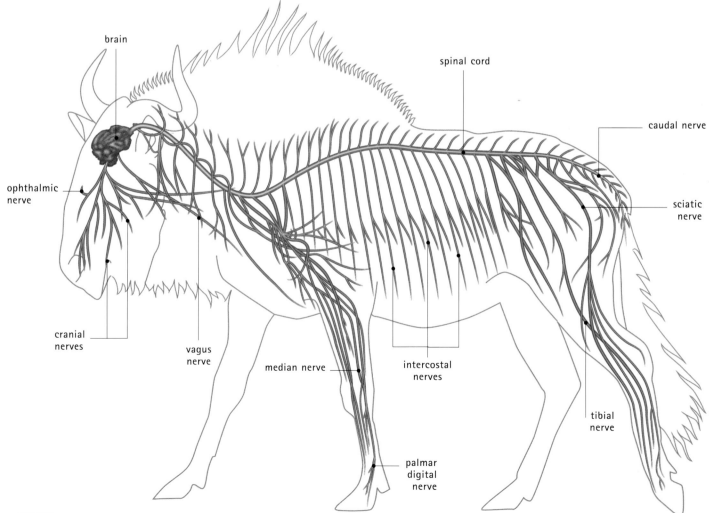

brain

spinal cord

caudal nerve

ophthalmic nerve

sciatic nerve

cranial nerves

vagus nerve

median nerve

intercostal nerves

tibial nerve

palmar digital nerve

nervous hops to one side as though starting to run away from the potential scene of conflict. Many of the ritualized displays performed by other ungulates appear to have evolved from similarly conflicting urges.

The eyes are on the sides of the head and have horizontally long pupils—an arrangement that provides the best possible range of view. A wildebeest can detect movement anywhere except directly behind its head. The sense of hearing is good, and the ears are able to swivel forward and backward to follow sounds.

A wildebeest's sense of smell, or olfaction, is also excellent. Scent is highly important in communication, and individual wildebeests respond to the smell of secretions from the various scent glands of other animals in the herd. The scent produced by the pedal gland in the lower leg is important in keeping herds together: a lone individual can use the herd's scent trail to follow and rejoin the herd. Wildebeests also smell the air to detect danger, often pausing to sniff when they are anxious. As well as the traditional smelling organ in the

IN FOCUS

Sleepless nights

Ruminant ungulates usually do not sleep deeply. They probably compensate for the lack of proper sleep with extended periods of drowsiness, usually associated with bouts of chewing the cud. Interestingly, the patterns of brain activity of a ruminating cow are similar to those of a nonruminant animal during normal sleep. However, ruminants keep their eyes open when chewing the cud, and they are aware of their surroundings. In this drowsy state, it seems they are able to gain some of the benefits of proper sleep without having to "switch off" completely. There are two very good reasons for maintaining some level of alertness: first, as prey animals, they need to stay alert to danger; second, as ruminants, they need to keep an upright posture even when resting, to help the complex stomach maintain its shape and reduce the chances of blockages or choking on regurgitated stomach contents.

nose, wildebeests have another chemosensory organ, located beneath the nostrils and opening in the roof of the mouth. This organ is called the vomeronasal organ, and it is particularly sensitive to pheromones, which are chemicals produced by animals for communication. To sample these chemicals, the wildebeest "sniffs" through the mouth and draws its upper lip back (a behavior called flehmen) to help draw in the pheromones.

▲ *Wildebeests have a large muzzle, and smell, or olfaction, is their most important sense. As well as organs of smell in the nose, wildebeests have a vomeronasal organ in the roof of the mouth to detect pheromones from other wildebeests.*

Circulatory and respiratory systems

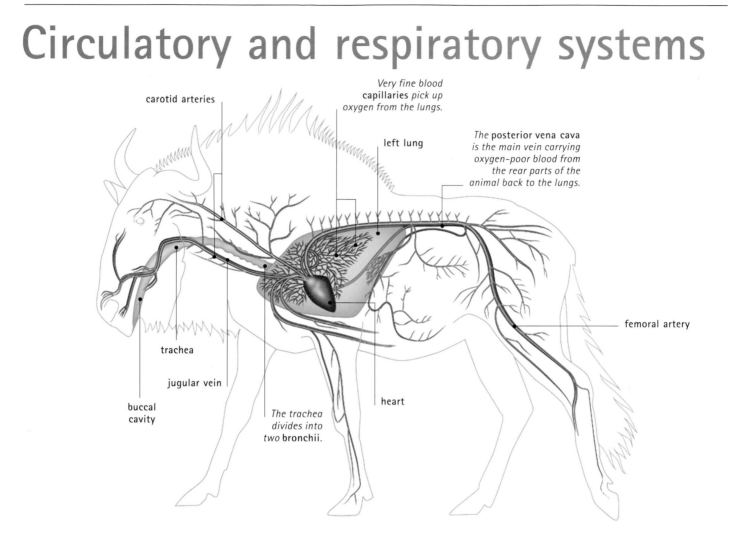

carotid arteries

Very fine blood
capillaries *pick up
oxygen from the lungs.*

left lung

The **posterior vena cava**
*is the main vein carrying
oxygen-poor blood from
the rear parts of the
animal back to the lungs.*

femoral artery

trachea

jugular vein

buccal
cavity

*The trachea
divides into
two* **bronchii.**

heart

A wildebeest's deep chest and large rib cage contain a pair of large, roughly pyramid-shape lungs. The lungs are made of moist, pink, spongy tissue containing millions of tiny interconnected air spaces called alveoli. The pink color is the result of a network of fine blood vessels called capillaries. Air is drawn into the lungs through the nose and mouth, which are connected to the lungs by a tube called the trachea, or windpipe. The trachea splits into two tubes called bronchi (singular, bronchus); each tube supplies one lung. Respiratory gases diffuse through the thin walls of the alveoli. Oxygen passes from the air into the blood, and carbon dioxide (a waste product of cellular respiration) passes from the blood to the air and is then expelled as the animal exhales.

Blood is pumped around the body by the heart. This organ lies slightly left of center at the base of the thoracic cavity, more or less between a wildebeest's forelegs. As in all mammals, the bovid heart has four chambers: a right atrium

and ventricle, which receive blood that has been around the body and pump it to the lungs for gas exchange; and a left atrium and ventricle, which receive oxygen-rich blood from the lungs and send it around the body. The chambers on both sides of the heart have a similar volume, but the left side of the heart is larger because the walls are thicker and more muscular.

▲ **Common** wildebeest
*A wildebeest has a four-chamber heart that
pumps blood, carried
in vessels, around the
body. Inhaled oxygen
reaches the blood via
the large lungs.*

IN FOCUS

Bloodsuckers

Wildebeests and other bovids are targeted by a variety of external parasites that suck their blood. Leeches, ticks, lice, fleas, tsetse flies, and mosquitoes latch on whenever they get the chance. These bloodsuckers do not take much blood and rarely cause serious problems themselves, but they can transmit infections and blood parasites such as those that cause malaria and sleeping sickness in humans. Wild bovids, including wildebeests, act as carriers of these dangerous diseases in Africa.

Digestive and excretory systems

Wildebeests belong to a group of grazing ruminants that eat large quantities of grass and herbage, washed down with plenty of water. They must drink at least every other day and are thus rarely found far from water. The multichamber stomach is large, muscular, and complex even by ruminant standards.

Forestomachs

The various compartments of the stomach are clearly defined, with muscular walls that thoroughly mix and churn the contents. The first three chambers—the rumen, reticulum, and omasum—are collectively called the forestomachs. They are lined with the same kind of tissue: simple squamous epithelium. However, beneath the epithelium are distinctive structures that help distinguish the different chambers and give clues to their function. The main job of the forestomachs is to act as holding areas for ingested food. Grass lingers for many hours in these three chambers, giving plenty of time for bacteria, which live mainly in the rumen, to go to work on the tough cellulose in the plant cell walls, breaking it down into more easily digestible sugars. The rumen is by far the largest of the stomach chambers, and it is subdivided by muscular folds into four pouches: the dorsal, ventral, caudodorsal, and caudoventral sacs.

Chewing cud

Wildebeests graze for long periods, giving each mouthful of food a cursory chew before swallowing until the rumen is full. They then spend several hours resting and chewing cud.

▼ *Large herds of wildebeests roam the savanna in search of food. They need to eat plenty of fresh grass and also have to drink water frequently.*

Chewing cud involves regurgitating food, one mouthful at a time, then chewing it again, this time very thoroughly. The chewing mixes saliva and millions of cellulose-digesting bacteria in with the food. By the time it is swallowed a second time, the grass is reduced to a moist pulp in which the bacteria are already at work breaking down the meal. Most of the bacteria in each mouthful are digested themselves when the food passes on, but those retained in the rumen reproduce at a great rate to compensate.

Wall structure

The second chamber of the forestomachs is the reticulum. There is no obvious separation between the rumen and the reticulum; ingested food moves easily from one chamber to the other and back again. The distinction has more to do with the structure of the chambers' walls. The lining of the rumen is covered with thousands of small cone-shape protuberances called papillae, whereas the internal walls of the reticulum are covered with small folds, or creases, that link up to form a polygonal network reminiscent of a honeycomb. In both the rumen and the reticulum, the walls are richly supplied with blood vessels, and some of the first products of digestion (mostly fatty acids) are absorbed

directly through the lining before reaching the true stomach (abomasum).

The reticulum is connected to the third stomach chamber, a small spherical pouch called the omasum. In fresh grass grazers such as wildebeests, the omasum is particularly well developed. It has a complex structure and serves as a kind of strainer though which large fragments of food cannot pass. They are regurgitated and chewed again. Like the rumen and reticulum, the omasum also provides an absorptive surface: its walls are highly permeable to water and certain nutrients released during the early stages of digestion. The lining of the omasum is convoluted into many thin folds, which look similar to the pages of a book. The folds greatly increase the internal surface area of the omasum, which despite being the smallest chamber has about one-third the total internal surface area of the forestomachs.

The final chamber is the abomasum; this part is the true stomach. It is equivalent to the single stomach of nonruminant mammals and has a similar structure. The inner walls are lined with glandular epithelial cells that secrete stomach acid and other digestive juices.

Having passed through the true stomach, food enters the intestines by way of the

▼ GRAZERS AND BROWSERS
This group of antelope shows—from left to right—a bushbuck, two four-horned antelope, an eland, and a nyala. Apart from the four-horned antelope, which live in Asia, all these antelope are found in Africa. They graze or browse on a wide variety of vegetation, including shoots, twigs, herbs, and fresh grass.

duodenum. The small intestine is long and convoluted and leads to the large intestine, which ends in the rectum. By the time food reaches the rectum, it is virtually all indigestible waste. This is then excreted from the anus.

The stomach of a true ruminant such as a wildebeest takes up about three-quarters of the abdominal cavity. Most of its volume is taken up by the rumen, where bacteria break down tough plant matter. In a horse, the stomach is relatively small, and the bacterial fermentation of plant material takes place in the hindgut instead. In carnivores and omnivores there is less need for predigestion with bacteria, so animals like bears have a relatively simple stomach. The camel is a pseudoruminant: its stomach has three chambers, in which a limited amount of breakdown of plant matter by bacteria takes place.

Removal of wastes

Like all other mammals, wildebeests have two kidneys that filter metabolic wastes and water out of the blood. In each kidney, the waste substances, which are dissolved in water to form urine, pass down a duct called a ureter into the hollow, muscular bladder, in which the urine is stored. When the bladder becomes full, the urine is expelled from the body through a duct called the urethra.

Success on the savanna

Fossil and DNA evidence suggests that the bovid family first appeared in Europe and Africa during the Miocene epoch, about 22 million years ago. The family underwent a sudden, very dramatic increase in numbers and diversity, which scientists think was triggered by changes in climate and vegetation type, particularly the appearance of vast areas of African grassland or savanna. Grass provides less nutrition than many other types of plant materials, but it became available in such vast quantities that within a relatively short period of time it provided the basis of one of the world's great ecosystems.

COMPARATIVE ANATOMY

Stomach complexity

Animals cannot break down cellulose, the tough carbohydrate that is the main constituent of plant cell walls. However, to extract sufficient nutrients from plants, many herbivores, especially ruminants such as the giraffe and wildebeests, are foregut fermenters—they have a complex multichamber stomach that houses bacteria that can break down cellulose into digestible sugars. Other herbivores, such as horses and zebras, have bacteria in their large intestine that can break down cellulose, and these animals are called hindgut fermenters.

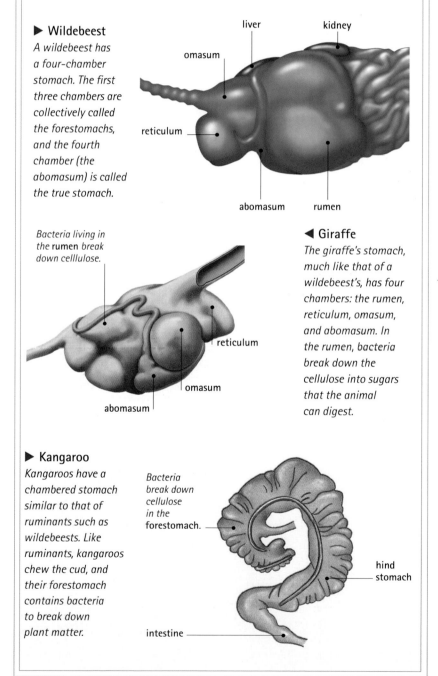

▶ **Wildebeest**
A wildebeest has a four-chamber stomach. The first three chambers are collectively called the forestomachs, and the fourth chamber (the abomasum) is called the true stomach.

liver   kidney
omasum
reticulum
abomasum   rumen

Bacteria living in the **rumen** *break down celllulose.*

reticulum

omasum
abomasum

◀ **Giraffe**
The giraffe's stomach, much like that of a wildebeest's, has four chambers: the rumen, reticulum, omasum, and abomasum. In the rumen, bacteria break down the cellulose into sugars that the animal can digest.

▶ **Kangaroo**
Kangaroos have a chambered stomach similar to that of ruminants such as wildebeests. Like ruminants, kangaroos chew the cud, and their forestomach contains bacteria to break down plant matter.

Bacteria break down cellulose in the forestomach.

hind stomach

intestine

Reproductive system

CONNECTIONS

COMPARE the large, well-developed young of wildebeests with the tiny embryonic joey produced by the *KANGAROO*. Young wildebeests need to be up and running as soon as possible after being born, to avoid predation. However, newborn kangaroos grow and develop further in the mother's pouch, where they are protected for up to 10 months.

Wildebeests are highly gregarious and social animals. Herds typically number 20 or 30 individuals, but at certain times of the year many herds may come together to take advantage of good feeding conditions or to make mass migrations, for which the species is renowned. Such superherds may comprise tens of thousands of animals.

Bachelor herds

A typical herd of wildebeests contains mostly females and their young. Young males are driven away from the nursery herd at one year of age, but young females remain part of the herd. The ejected males join bachelor herds and may spend three or four years growing and maturing in the company of other males. After that, the strongest leave the herd and take on a territory, which they defend from rivals by using intimidation and sometimes by ramming horns. When the mating season arrives, the dominant males lay claim to the females in their territory.

At the time of mating, the female uterus is a compact organ with two distinct "horns" into which the tubes connecting the ovaries to the uterus open. Having mated, females are pregnant for almost nine months. They bear just one calf at a time. The embryo develops within one or the other of the uterine horns (in cattle, for some reason, it is more often the right than the left) and the uterus expands to accommodate it. Close to full term, the uterus is so large that it displaces much of the female's intestine and puts pressure on the rumen and the diaphragm.

The delivery of a newborn wildebeest is relatively rapid. The female is usually in active labor for only about an hour, and the newborn is amazingly precocious. It will stagger to its feet within five minutes of birth, and after just one day can run as fast as an adult wildebeest. This is a crucial adaptation because any animal that cannot keep pace with the rest of the herd will be picked off quickly by predators, such as lions and hyenas.

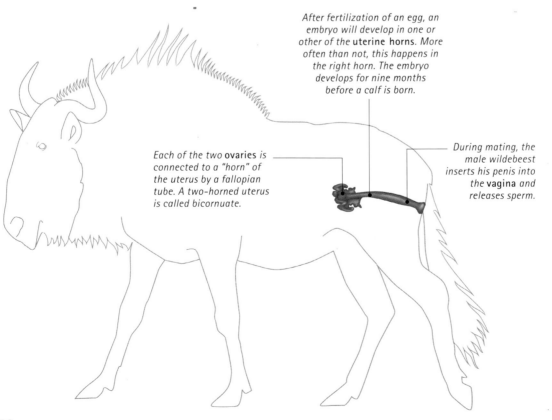

After fertilization of an egg, an embryo will develop in one or other of the **uterine horns**. More often than not, this happens in the right horn. The embryo develops for nine months before a calf is born.

Each of the two **ovaries** is connected to a "horn" of the uterus by a fallopian tube. A two-horned uterus is called bicornuate.

During mating, the male wildebeest inserts his penis into the **vagina** and releases sperm.

◀ Common wildebeest
The female reproductive system consists of a pair of egg-producing ovaries and a uterus with two horns, called a bicornuate uterus. The male reproductive system consists of two sperm-producing testes and a penis.

▶ *After a year, young males are expelled from the herd. They join bachelor herds, in which they may stay for a few years. The stronger males break away and establish a territory, which they defend from other males by ramming horns.*

The placenta: A temporary organ

As with all placental mammals, the final stage in the birthing process is the expulsion of the placenta, called the afterbirth. The placenta is a temporary organ that forms during gestation. It allows essential nutrients and the vital gas oxygen to pass from the mother's bloodstream to that of the fetus and fetal wastes to move in the opposite direction. Once a young wildebeest is born, the placenta is no longer needed and is shed from the uterine wall quickly, usually within a period of minutes.

Like all young mammals, a newborn calf is fed exclusively on milk, which is a nutritious solution of fats, proteins, and sugars. Milk is secreted from glands in the mother's abdominal wall and delivered to the calf by way of teats, in response to sucking.

The first drink of milk is very important. It marks the beginning of a special bond between mother and calf. During the first few minutes after birth they learn to recognize each other's

scent, and the calf imprints on its mother. From now on the calf will instinctively follow her and be able to pick her out from the rest of the adults in the herd. Equally important is the milk itself. In all mammals, the first milk produced

▼ *Female wildebeests give birth, or calve, very quickly, within the protection of a herd and not in a secluded place.*

Synchronized breeding

Wildebeests have a very short breeding season. Between 80 and 90 percent of all the pregnant females in a herd deliver their calves within the same three-week period, at the start of the rainy season, which occurs from March to May. The result is a short period of very rich pickings for predators, such as big cats, and inevitably many young wildebeests are killed and eaten—but fewer overall than would be lost if the wildebeests bred year-round. By synchronizing births so closely, the wildebeests create a glut of potential food for predators. There are so many young that even working overtime the predators can kill only a small percentage of the young calves.

▼ *Female wildebeests give birth to a single calf. The calf stays close to its mother until she gives birth again the following year. At that time, a female calf remains in the herd; if the calf is male, it is driven away from the herd.*

after birth is very special. It is called colostrum, and it contains a host of antibodies (protective proteins) from the mother's body. This milk provides a kind of vaccination for the newborn, which helps make it much less vulnerable to everyday infections. Normally, these antibodies would not survive the digestive process, but in the first few hours after birth it is possible for the antibodies to be absorbed through the immature wall of the calf's stomach and into the bloodstream.

From milk to grass

Starting life on an exclusively milk diet has important implications for the development of the digestive system, and the foregut of newborn bovids is very different from that of an adult. In fact, when a wildebeest is born, the abomasum is the only fully developed chamber of the stomach. This chamber is equivalent to the true stomach of nonruminant animals. The other chambers do not yet need to function, because the constituents of milk—proteins, fats, and a sugar called lactose—do not require bacterial fermentation in order to be broken down for absorption. The pink glandular lining of the abomasum secretes digestive juices that do the job perfectly well. In a newborn calf, the

rumen, reticulum, and omasum—the chambers called the forestomachs—are very small. In fact, the abomasum is bigger than the others combined. It is only when the calf begins to show an interest in grass—perhaps chewing a little in imitation of its mother at a few weeks of age, that dramatic changes begin to take place. The chambers begin a rapid growth spurt. Weaning is usually complete by five months, but the youngster may continue to drink some milk after that for comfort.

JOHN WOODWARD

▲ *Wildebeest calves suckle until they are about five months old, after which time their diet is almost exclusively fresh grass and water.*

FURTHER READING AND RESEARCH

Macdonald, David W. 2006. *The Encyclopedia of Mammals.* Facts On File: New York.

Vrba, Elisabeth S., and George B. Schaller (eds.). 2000. *Antelopes, Deer, and Relatives: Fossil Record, Behavioral Ecology, Systematics, and Conservation.* Yale University Press: New Haven, CT.

The Hall of Mammals:

www.ucmp.berkeley.edu/mammal/mammal.html

CLOSE-UP

Eating the afterbirth

More often than not, female animals, even herbivorous ones, consume the afterbirth in an effort to replenish some of their body's lost reserves. For wildebeests, there is a further advantage to disposing of the placenta in this way. The organ contains a large quantity of blood, the scent of which will very quickly attract predators or scavengers, especially hyenas and jackals. Given half a chance, these carnivores will devour not only the afterbirth but also the newborn calf. Thus it is in the mother's interest to dispose of the placenta as quickly as possible. Female wildebeests also sometimes eat the dung of their young offspring. Presumably this behavior, too, is to avoid leaving telltale signs for predators.

Wolf

ORDER: Carnivora   FAMILY: Canidae   GENUS: *Canis*

There are two species of true wolves in the genus *Canis*: the gray, or timber, wolf; and the red wolf. Some biologists consider the red wolf to be a hybrid and not a species in its own right. With a natural range taking in most terrestrial regions of the Northern Hemisphere, the gray wolf is one of the world's most widespread mammals. It is a superb predator, built for athleticism and endurance. Wolves have sharp vision and a phenomenally acute sense of smell. The biggest secret of their success, however, is teamwork—wolves are among the most social of all mammals.

Anatomy and taxonomy
Scientists categorize all organisms into taxonomic groups based on anatomical, biochemical, and genetic similarities and differences.

● **Animals** Wolves, like other animals, are multicellular and fuel their body by eating organic material (food) from other organisms. Animals differ from other multicellular life-forms in their ability to move from one place to another (in most cases, using muscles). Animals are sensitive to stimuli such as touch, light, and various chemicals.

● **Chordates** At some time in its life cycle, a chordate has a stiff, dorsal (back) supporting rod called the notochord that runs all or most of the length of the body.

● **Vertebrates** In vertebrates, the notochord develops into a backbone (spine or vertebral column) made up of units called vertebrae. The vertebrate muscular system that moves the head, trunk, and limbs consists primarily of muscles arranged like a mirror image on either side of the backbone (bilateral symmetry about the skeletal axis).

● **Mammals** These vertebrate animals are warm-blooded and have hair made of keratin. Females have mammary glands that produce milk to feed their offspring. In mammals, the lower jaw is a single bone, the dentary, whereas in other vertebrates it is several fused bones. A mammal's inner ear contains three small bones (ear ossicles). Mature mammalian red blood cells lack a nucleus;

▼ *Wolves are mammals in the order Carnivora and family Canidae. There are two species of true wolves—gray and red wolves—both in the genus* Canis. *This genus also includes the dingo, domestic dog, coyote, Ethiopian wolf, and jackals. The maned wolf is placed in a different genus,* Chrysocyon.

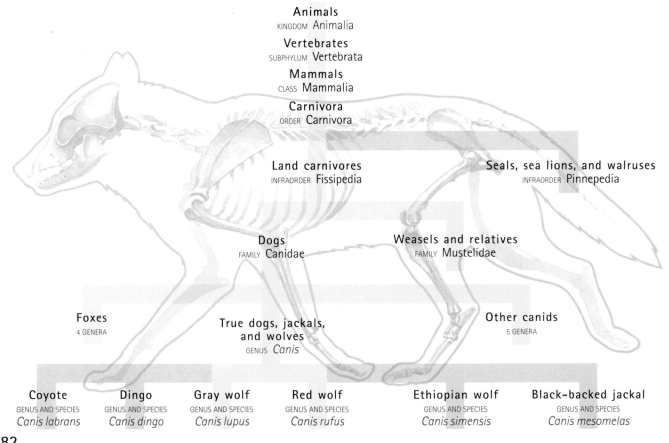

Animals
KINGDOM Animalia

Vertebrates
SUBPHYLUM Vertebrata

Mammals
CLASS Mammalia

Carnivora
ORDER Carnivora

Land carnivores
INFRAORDER Fissipedia

Seals, sea lions, and walruses
INFRAORDER Pinnepedia

Dogs
FAMILY Canidae

Weasels and relatives
FAMILY Mustelidae

Foxes
4 GENERA

True dogs, jackals, and wolves
GENUS *Canis*

Other canids
5 GENERA

Coyote
GENUS AND SPECIES
Canis labrans

Dingo
GENUS AND SPECIES
Canis dingo

Gray wolf
GENUS AND SPECIES
Canis lupus

Red wolf
GENUS AND SPECIES
Canis rufus

Ethiopian wolf
GENUS AND SPECIES
Canis simensis

Black-backed jackal
GENUS AND SPECIES
Canis mesomelas

all other vertebrates have red blood cells that contain a nucleus. Placental mammals nourish their unborn young through a placenta, a temporary organ that forms in the mother's uterus during pregnancy.

● **Carnivores** The word *carnivore* can be used to describe any animal that eats meat, but it applies more specifically to members of the mammalian order Carnivora. Members of this group include dogs, cats, bears, raccoons, mustelids, civets, hyenas, and their close relatives. Most members of

▲ *These gray wolves are howling—a form of communication that helps regroup a dispersed pack, signifies the beginning of a hunt, or tells other packs of wolves to keep off their territory.*

the group eat meat almost exclusively, but some have a mixed diet. One member of the group, the giant panda, eats only plants. Carnivores have cheek teeth called carnassials, which are specialized for slicing flesh. Another characteristic is that males have a penis bone, or baculum, which supports the penis and prolongs mating.

EXTERNAL ANATOMY Wolves are quadrupeds with a narrow body, deep chest, and long bushy tail. The head is large with a slender pointed snout and large ears. The fur is shaggy. *See pages 1385–1389.*

SKELETAL SYSTEM Wolves have long legs with separate bones in the forelimbs and no collarbone. The skull bears a ridge called the sagittal crest for the attachment of the powerful jaw muscles. *See pages 1390–1391.*

MUSCULAR SYSTEM Wolves are lithe, athletic animals. The neck, shoulders, and hips are muscular. Smaller muscles in the face and tail allow the fine movements that are important in visual communication. *See pages 1392–1393.*

NERVOUS SYSTEM Wolves are intelligent, with excellent vision and hearing and a phenomenally acute sense of smell. The vomeronasal organ in the roof of the mouth

provides wolves with an additional olfactory sense. *See pages 1394–1396.*

CIRCULATORY AND RESPIRATORY SYSTEMS Wolves are warm-blooded with a typical mammalian circulation. The lungs are large, and the larynx and vocal cords can produce a range of vocalizations. *See page 1397.*

DIGESTIVE AND EXCRETORY SYSTEMS Wolves are committed carnivores with carnassial teeth suited to slicing up meat. The intestine is relatively short and simple. Metabolic waste is removed from the blood by a pair of efficient kidneys. *See pages 1398–1399.*

REPRODUCTIVE SYSTEM Female wolves bear litters of young and nourish them with milk from 8 or 10 nipples. Males have a baculum (penis bone). Wolves are social and live in cooperative family groups. *See pages 1400–1403.*

FEATURED SYSTEMS

▲ *Despite their name, gray wolves have coats varying in color from near white to black with a yellow, red, or brown tinge. Only a small percentage of gray wolves—less than 5 percent—are black.*

▲ *Compared with the gray wolf, the red wolf is smaller in stature and has relatively longer legs, a narrower body, and larger ears. In addition, its coat is shorter and redder than that of a gray wolf.*

● **Dogs** Members of the family Canidae are generally easy to recognize. They are fully quadrupedal (unlike bears or mongooses, which often stand on their two hind legs). They have a narrow body and a deep chest, and most are long-legged with a bushy tail. All except one species have four digits on each hind foot and five on each front foot. One of the front digits is a vestigial (evolutionary leftover, or remnant) that does not reach the ground, called a dewclaw. The African wild dog lacks dewclaws. All dogs have blunt claws that are used for traction when running. The main weapons for hunting and combat are the teeth. The snout is characteristically long, and in wild species the ears are usually large and erect. There are 34 species of wild canids and several hundred breeds of domestic dogs,

which far outstrip their wild relatives in variety of forms. However, all types of domestic dogs are considered a single species. Most dogs are social and use vocal communication, which includes barks, howls, growls, and whines.

● **True dogs, jackals, and wolves** Members of the genus *Canis* are distinguished by long legs and a bushy tail, which is not as thick or rounded as that of the foxes (genus *Vulpes*). The pupils of the eyes remain round in bright light, whereas those of foxes tend to look oval. Members of the genus *Canis* are all social, living in family groups that may coalesce into packs of 30 or more usually related animals. Apart from gray and red wolves, other members of the genus include the coyote, dingo, and jackals.

External anatomy

The gray wolf is the archetypal wolf—a long-legged, shaggy-coated, rangy-looking dog, with a large head, a pointed snout, a deep chest, a narrow trunk, and a long, bushy tail. As in many animals with a very widespread distribution, there are geographical variations in appearance among wolves from different parts of a species' range. Local circumstances have created these differences. For example, the most obvious difference between a Canadian timber wolf and a Mexican gray wolf is size. The northern variety is suited to the crushing cold of subarctic winters, when large size is a definite advantage. Canadian and Alaskan gray wolves are the biggest in the world, with a large male weighing up to 180 pounds (80 kg).

Compare wolves of this size with the gray wolves from Mexico or Egypt—still the same species—where full-grown adults may weigh less than 44 pounds (20 kg). Not surprisingly, tundra-dwelling wolves also grow a much heavier winter coat (pelage) than gray wolves living in warmer climates, and this pelage makes them look even move impressive.

As their name suggests, gray wolves are generally a shade of gray, though coat color varies from off-white to black and may be tinged with red, brown, or yellow. The fur is thickest on the back and shoulders, where it forms a rough mane in some animals. The fur is thinnest on the belly.

The coat has two types of hairs. Primary hairs, or guard hairs, are long and pigmented, with a long cylindrical shaft that tapers to a point at the tip. Guard hairs grow from follicles in the outer layer of the skin. These follicles are usually arranged in tight rows all over the wolf's body, except on the pads of the feet and the tip of the nose. The guard hairs are coarse and give a wolf its shaggy appearance. The surface of each hair is made of many slightly overlapping scales. These give the shaft a distinct feel: if you were to slide your fingers along a wolf's hair, it would feel much smoother from base to tip than the other way. This "nap" is significant because when a wolf gets wet, water tends run out of the coat, away from the skin, rather than soaking in.

◀ The maned wolf is the largest canid in South America, with a shoulder height of almost 39 inches (1 m). This cousin of the true wolves has a long, golden-red coat with a black mane of hairs that stand erect.

The second type of hairs, called secondary hairs, or awns, form a dense underlayer of fur. Awns are very fine, soft hairs, and there are up to several dozen for every guard hair. Awns provide insulation—by trapping a layer of air close to the skin, they help the wolf keep warm. Wolves in warm climates have much thinner underfur than those in the far north. Both types of hairs are kept slightly greasy by secretions from tiny glands in the skin. The grease helps condition the fur and makes it resistant to water.

As well as the primary and secondary hairs of the coat, wolves have two additional types of hairs. The upper lid of each eye bears a row of eyelashes, or cilia, which protect the surface of the eye from particles of dust or debris and from drops of water. There are no lashes on the lower eyelid. Wolves also have many long, sensory hairs or whiskers (also called

▲ *Depending on the region, adult gray wolves weigh between 44 to 180 pounds (20–80 kg). In general, they are larger than red wolves and considerably larger than most types of domestic dogs.*

▶ *Gray wolf*
The coat is most commonly gray or yellowish brown but may also be other shades or have a red tinge. The body shape is streamlined with long legs and a bushy tail. The head has a narrow snout, or muzzle, and pointed ears.

The eyes are large and round and face forward. Eye color varies but is usually a shade of gold, brown, or even blue. The whites of the eye are often visible.

The ears are large and pointed. Hearing is excellent—wolves can hear very faint sounds made by prey from a considerable distance. Wolves rely more on hearing than vision for hunting.

The canine teeth—from which canids take their name—are used for tearing flesh.

The nose is hairless and black, brown, or sometimes pink. The surface of the nose has tiny fissures. The nostrils open under curving flaps. Smell is a wolf's most important sense for tracking prey and recognizing other wolves.

The legs are long and slim. Wolves walk with a trotting pace and leave a single line of paw prints. Both forepaws and hind paws have four functional toes with claws. The forepaws also have a fifth nonfunctional claw (dewclaw).

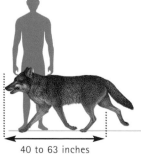

40 to 63 inches
(1–1.6 m)

vibrissae). Whiskers are deeper-rooted than normal hairs and their follicles are richly supplied with blood and nerves. They are located strategically around the wolf's body, mostly on the head. The whiskers are located in rows on the upper lip and in tufts on the lower lip, chin, throat, and cheeks, and above the eyes. Whiskers provide the wolf with excellent spatial awareness when it is moving about in the dark. However, canids are generally less reliant on their whiskers than cats are. Canids spend less time moving about in cluttered environments (such as among the branches of trees) and are often active during the day when it is light enough to see.

COMPARATIVE ANATOMY

Red fox and gray wolf

There are 10 species of foxes, and one, the red fox (*Vulpes vulpes*), rivals the gray wolf as the world's most widespread carnivore. Foxlike dogs are referred to as "vulpine" species. The differences between foxes and wolves generally have to do with scale and proportion. Foxes are smaller and have shorter legs. The tail is relatively large in fox species and often very bushy (and called a brush). Foxes also usually have very large ears relative to the size of the head, and the pupils of their eyes contract to ovals in bright light. Foxes produce a particularly strong musky scent from the caudal (tail) glands.

The **tail** is long, bushy, and drooping. It varies in length from 12 to 20 inches (30–50 cm). Wolves and dogs wag their tail to communicate mood.

The **body** is streamlined and built for speed. The head is narrow and pointed, the body is slender with smooth fur, and the tail is long and pointed.

The **fur** has two types of hairs. Soft, thick underfur keeps a wolf warm and dry. Long guard hairs keep snow and water out. The blotchy markings of the coat match the habitat, camouflaging the wolf as it stalks prey.

The **claws** cannot be retracted, unlike those of most species of cats.

Red wolf

The North American red wolf, *Canis rufus*, was once thought to be the world's rarest species of dog. In 1975, when its numbers had shrunk to an all-time low of just a few dozen, conservationists took the drastic step of taking the entire population into captivity so that a captive-breeding program could be started. This program was very successful and red wolves have since been reintroduced into the wild on a number of refuges in the United States. However, wolf introductions are always controversial, and in the early 1990s the antiwolf lobby seized on the piece of genetic research that suggested the red wolf may in fact be a hybrid between the gray wolf and the coyote and not a species in its own right at all. Opinion remains divided, and meanwhile the future of the red wolf hangs in the balance.

The wolf's head is large relative to the size of its body. The domed section accommodates a relatively large brain, and the long, tapering snout contains phenomenally sensitive olfactory (smelling) equipment. The hair on the face is shorter and sleeker than elsewhere on the body. Not only does this make it easier to keep the face free of dirt during feeding; it also makes changes in facial expression more obvious. The external part of the ear (the auricle) consists of a large, triangular flap, usually held erect, though in some wolves it may be slightly floppy or torn. The ear can be rotated slightly, allowing the wolf to focus directly on the source of a sound. Movement of the ears also adds to the wolf's repertoire of facial expressions.

The eyes are large, round, and predominantly forward-facing. However, they also bulge slightly to the sides of the head, giving some peripheral vision. The color of the eyes varies but is usually some shade of gold or brown. Blue-eyed wolves are not uncommon. Wolves are among relatively few animals in which the whites of the eye are often visible. This is thought to be a feature that enhances communication in social species—making it easier for other members of the group to see where an individual is looking. (One reason human eyes are so expressive is that they show a lot of white around the iris, the colored part of the eyes.) The wolf's nose is hairless and usually black or brown—or occasionally pink. Its surface is covered in tiny fissures, and the nostrils open under curving flaps to either side.

Domestic varieties

Domestic dogs are direct descendants of wolves. They belong to the same species and share the same scientific name, *Canis lupus*; domestic dogs are *Canis lupus familiaris*. The connection is obvious in some breeds —the husky is effectively a tame wolf—and less so in others. It is difficult to imagine how a Pekingese or toy poodle can be first cousin to a wolf. Wolves were probably first tamed over 100,000 years ago, and some biologists think that different gray wolf subspecies gave rise to the main groups of modern domestic stock. So working dogs like spaniels and setters have an ancestry different from terriers, whereas yet another wolf subspecies was the starting point for the bulldog–boxer group.

▶ *All domestic dogs, including this standard poodle, are descendants of wild wolves. There are about 400 breeds of dogs, ranging in size from the tiny Chihuahua to the tall Irish wolfhound and Great Dane.*

Beneath the fur, the wolf's skin is usually pink with an extensive and variable mottling of dark brown to bluish black pigment, similar to that seen on domestic dogs. In addition to the glands that produce oils to condition the fur, there are a number of other glands than open directly onto the skin. Like all mammals, wolves have mammary glands on their underside, which in adult females produce milk to nourish the young. Most canids have five pairs of nipples, each of which contains the openings of about a dozen tiny ducts leading from the mammary glands. Another concentration of glands occurs on the top of the base of the tail. These caudal glands produce a scent unique to each wolf, but the scent of wolves is nowhere near as powerful as that of foxes.

Sweat-producing glands are notable by their absence over most of the body surface. Dogs do not sweat. Their thick coat prevents moving air from making contact with the skin, so sweating would be a much less effective means of cooling than it is for humans. Thus the animals must find alternative means of cooling. The long, lolling, pink tongue is an excellent cooling surface. It is permanently moist and

▼ MAKING TRACKS
Wolves have fives toes on their forefeet and four toes on their hind feet. One of the toes of the forefeet—the dewclaw—is vestigial and does not make contact with the ground. The claws of the toes cannot retract and are visible in footprints.

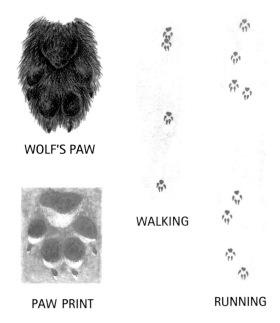

WOLF'S PAW

WALKING

PAW PRINT

RUNNING

CLOSE-UP

Express yourself

A large part of the teamwork that allows wolves to hunt successfully in packs is due to communication. It is no coincidence that the face of an average dog is considerably more mobile than that of a cat—cats are mostly solitary, whereas dogs are inherently social. Wolves and dogs can raise their eyebrows to show interest or alertness; gape in a happy, relaxed "laughing" face; or wrinkle the nose and curl the lip to bare the teeth to signal a threat or fear. This limited range of expression serves to emphasize other body language such as posture, tail position, and the raising and lowering of hair on the hackles (back of the neck).

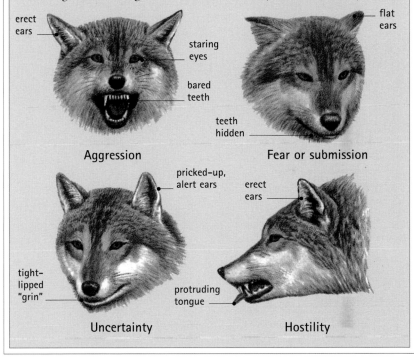

erect ears

staring eyes

bared teeth

flat ears

teeth hidden

Aggression

Fear or submission

pricked-up, alert ears

erect ears

tight-lipped "grin"

protruding tongue

Uncertainty

Hostility

well supplied with blood vessels, and the animal's breathing generates a continuous flow of air over its surface. A panting dog is not necessarily tired or out of breath; it may be cooling itself.

Wolves have four functional toes on each foot, and the front feet have a vestigial hallux (first digit) that forms the dewclaw. Each toe has a well-developed pad, with very thick callused skin covered by small bumps, or papillae, which help provide a grip like the tread of a shoe or car tire. Each pad has a wad of fatty tissue inside it to provide cushioning, and the whole structure has a rich supply of blood vessels. This helps prevent frostbite in the cold. In warm conditions, the foot pads are the only place from which the wolf can sweat.

Skeletal system

The wolf skeleton is that of a long-distance endurance athlete: strong but not heavy, with long, slender limbs and a deep chest to accommodate large lungs. Dogs are cursorial, or running, mammals, and their skeleton is structured accordingly. They have a digitigrade stance: that is, they stand and run on their toes—not on the very tips, like an ungulate (hoofed animal), but on the flat of the last joint in each digit. The bones inside the toes are short compared with those of a long-fingered mammal like a primate, sloth, or bat, but the bones of the legs (metacarpals in the forelegs and metatarsals in the back legs) are long. This adds considerably to the overall length of the leg and allows a wolf to take much longer strides than it would if it had a flat-footed, or plantigrade, stance like that of a bear or a human.

Protective skull

The skull is arguably the most important part of the skeleton because it protects the brain. As in other vertebrates, the skull is made of several bones that become fused as the animal grows and develops. In contrast, the lower jaw is just one bone, the dentary, which is a uniquely mammalian feature. In other vertebrates, such as fish and reptiles, the lower jaw includes additional bones. Thus the mammalian lower jaw is very strong—the hardest and densest bone in the body. The places where the skull bones fuse are marked by hairline fissures. Small holes in the front and sides of the skull allow nerves and blood vessels to pass from the brain to the muscles of the face, for example. In addition to the large braincase, the skull has a large cavity in the snout, which houses the olfactory organs—which are concerned with the sense of smell—and two large, round eye sockets, called orbits.

The skull articulates with the rest of the skeleton by way of a single joint between the base of the skull and the first cervical (neck)

▼ **Gray wolf**
The bones of wolves are strong, giving them the power to bring down large prey such as caribou. The narrow shoulder blades and long limb bones limit flexibility but make wolves efficient runners.

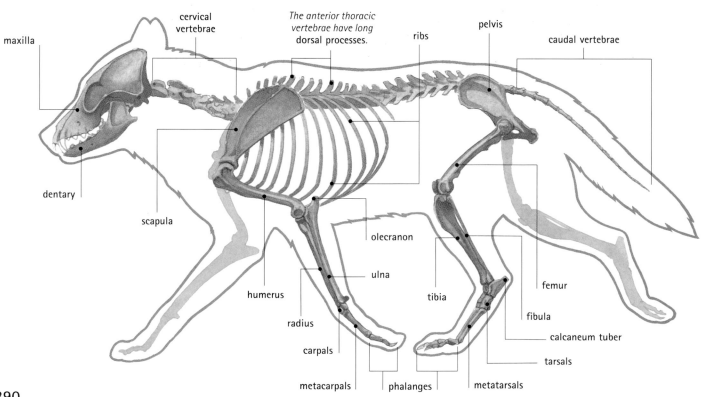

maxilla · cervical vertebrae · The anterior thoracic vertebrae have long dorsal processes. · ribs · pelvis · caudal vertebrae · dentary · scapula · olecranon · ulna · humerus · radius · carpals · metacarpals · phalanges · tibia · metatarsals · femur · fibula · calcaneum tuber · tarsals

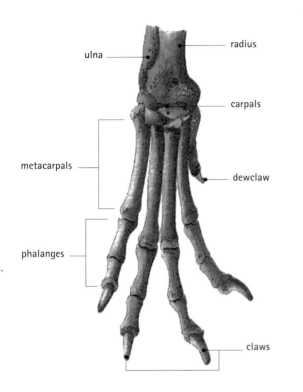

ulna — radius

carpals

metacarpals

dewclaw

phalanges

claws

▲ BONES OF THE FOREFOOT

The wrist bones of wolves are fused for extra strength. When the toes are splayed, a wolf can grip onto slippery, uneven, or steep surfaces. When closed, the toes form a strong paw.

vertebra, called the atlas. The rest of the wolf's spine is made of a further 6 cervical vertebrae; 13 thoracic (chest) vertebrae; 7 lumbar (lower back) vertebrae; 3 sacral vertebrae fused into a single bone, the sacrum, which supports the pelvis; and at least 9 caudal (tail) vertebrae. Each of the thoracic vertebrae supports a pair of long, gently curved ribs, which are connected to the breastbone, or sternum, by sections of elastic cartilage. Dogs lack a clavicle, or collarbone, connecting the bones of the forelimb with the sternum. The absence of the clavicle limits the movement of the shoulder joint to a single plane—forward and backward —but improves the efficiency of the running stride. Therefore, over millions of years of evolution, wolves have sacrificed a good deal of flexibility for speed.

▶ *The long limbs and streamlined body of the gray wolf makes it a fast runner, even in deep snow. When sprinting after prey, the gray wolf can reach speeds of up to 43 miles per hour (70 km/h).*

COMPARATIVE ANATOMY

Specialized jaws

Compare the skull of the gray wolf with the skulls of two other wild canids: the maned wolf and arctic fox. The diets of these other species are reflected in the structure of the jaws and teeth. The maned wolf eats small prey, such as mice, lizards, and insects. It also eats eggs and plant matter. Its jaws are suited to snapping up prey but not for cutting it up. The jaws are light and fast, with simplified peglike teeth for gripping an item of prey before it is swallowed whole. The arctic fox eats a lot of carrion. It cannot afford to be fussy and must make the most of any food it can find. Consequently, its jaws are stout and powerful waste-disposal tools capable of crunching up bones and other tough tissues.

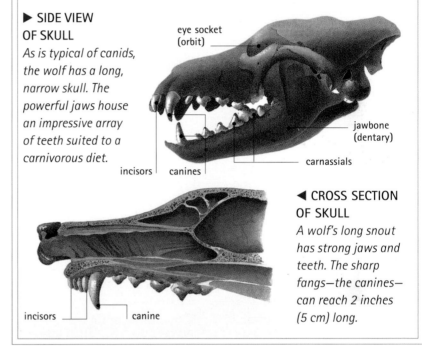

▶ SIDE VIEW OF SKULL
As is typical of canids, the wolf has a long, narrow skull. The powerful jaws house an impressive array of teeth suited to a carnivorous diet.

eye socket (orbit)

jawbone (dentary)

carnassials

incisors   canines

◀ CROSS SECTION OF SKULL
A wolf's long snout has strong jaws and teeth. The sharp fangs—the canines— can reach 2 inches (5 cm) long.

incisors   canine

Muscular system

Canine musculature is similar to that of most other carnivores and follows the same basic plan as that of most mammals. There are three types of muscles. Smooth muscle lines the walls of internal organs such as the intestine, bladder, uterus, and large blood vessels. It usually provides slow, low-energy contractions, and it does not tire. Smooth-muscle contraction is controlled by the autonomic nervous system (part of the peripheral nervous system), and the contractions are involuntary.

The second type, cardiac muscle, is closely related to smooth muscle. Its contractions are involuntary and tireless, keeping a wolf's heart beating at an average 120 beats per minute throughout its life.

The third type of muscle is skeletal muscle, which is also called striated muscle because under magnification, rows of microscopic fibers can be seen lining up to form striations, or stripes, in the tissue. Skeletal muscle is under voluntary control.

Under the skin, the first layer of muscle is cutaneous muscle. This allows the skin to quiver and twitch, and controls the lie of the fur—for example, raising the hackles when a wolf feels aggressive. The cutaneous layer is also able to accumulate fat when a wolf is well fed. Beneath the cutaneous muscle, the next layer, containing muscles such as the abdominal obliques, forms a taut sheath around the trunk and limbs. This helps keep the vital organs in place and prevents blood and lymph from pooling in the legs under the effects of gravity.

Deeper still lie the muscles that control posture and locomotion. They are arranged symmetrically within the body, and they act in

▼ Gray wolf

Wolves are sleek but also very muscular. Large trunk muscles drive the slender legs, allowing the animal to run fast and leap far. Strong neck muscles hold up the head, and powerful jaw muscles give wolves their ferocious bite.

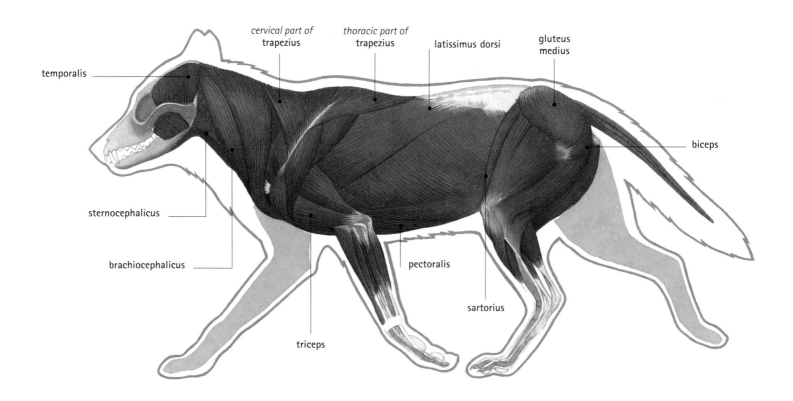

cervical part of
trapezius

thoracic part of
trapezius

latissimus dorsi

gluteus
medius

temporalis

biceps

sternocephalicus

brachiocephalicus

pectoralis

sartorius

triceps

antagonistic pairs to move parts of the skeleton to and fro: for every muscle that pulls a bone in one direction, there is another to move it back to its original position.

Powerful runner

The gray wolf can run at up to 43 miles per hour (70 km) over a short distance, and maintain a slightly slower pace for extended periods. It can travel continually for hours or days at a time and leap up to 16 feet (5 m) in a single bound. When hunting, wolves chase down their prey (usually deer) over a distance of a few hundred feet to several miles. Once overtaken, the prey is leaped at from the side and knocked to the ground. It usually takes the strength of several wolves to subdue a large deer.

IN FOCUS

Why do dogs wag their tail?

The muscles of the longissimus system continue into a wolf's tail and can raise the tail, lower it, and wag it from side to side. As every dog owner knows, the tail signals mood. It is carried high when a dog feels confident or aggressive and is tucked in when a dog is submissive or afraid. Loose, airy wagging is usually interpreted as a sign of happiness. However, behavioral scientists think that tail wagging begins as a sign of conflict, reflecting an issue—"Should I stay or should I go?"—in the dog's mind. Dogs are smart, communicative animals. They learn that wagging the tail often elicits a favorable response from other dogs and humans—like a human smile, which is also thought to have evolved from a sign of anxiety.

◀ As well as being able to sprint at great speed, wolves can jump high, leaping up to 16 feet (5 m). Their speed and agility make wolves supreme predators.

Nervous system

The nervous system is the hard wiring of the animal. Nerve cells, or neurons, are excitable cells that are able to transmit electrical and chemical signals to other cells and stimulate activities such as muscular contractions or the release of hormones. In the wolf, as in other mammals, the nervous system incorporates a main control center, the brain and spinal cord; and a complex branching network of peripheral nerves that carry signals to and from every part of the body. Not surprisingly, some organs are more heavily supplied with nerves than others. Neurons are not the only cells in the central nervous system. There are also backup cells, offering structural support and packaging.

Brainpower

The brain of the gray wolf weighs 4 to 5.6 ounces (120–150 g). This mass is about half as big again as the brain of a similar-size domestic dog. Domestic animals of all kinds (horses, sheep, cats, and others) nearly always have smaller brains than their wild ancestors and

relatives. The simple truth is that with humans taking care of their day-to-day needs, such as by providing food and territory, they do not need the same brain capacity. However, brain size is not necessarily a good indicator of intelligence, and experts are divided on whether or not wolves are actually smarter than dogs. Certainly, wolves are intelligent, highly emotional, and better equipped to cope with the challenges of life in the wild. They have a good memory and the ability to learn by association, just as dogs do.

Canids are relatively advanced vertebrates, with a highly evolved brain. The brain is divided into several distinct regions, with the oldest parts at the base, close to the top of the spinal cord. The medulla oblongata (myelencephalon) is the narrow neck of the

▼ Gray wolf
Although its structure is very similar, a wolf's brain is both larger and heavier than that of a domestic dog. Other features of the nervous system are virtually identical to those of dogs.

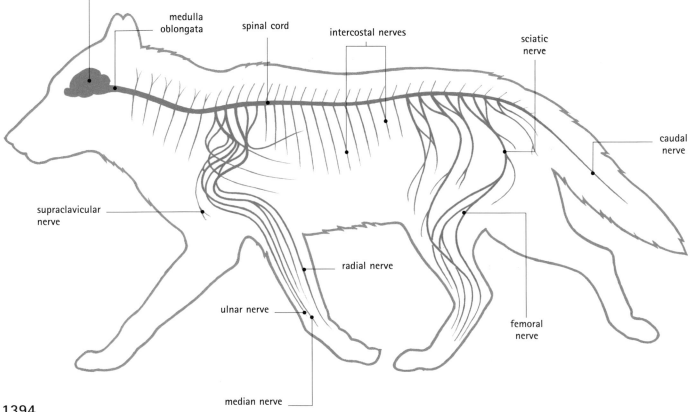

brain

medulla oblongata

spinal cord

intercostal nerves

sciatic nerve

caudal nerve

supraclavicular nerve

radial nerve

ulnar nerve

femoral nerve

median nerve

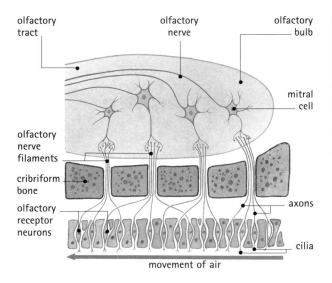

olfactory tract
olfactory nerve
olfactory bulb
mitral cell
olfactory nerve filaments
cribriform bone
olfactory receptor neurons
axons
cilia
movement of air

▲ VOMERONASAL ORGAN
Located in the floor of the nasal cavity, this organ, also called Jacobson's organ, is separate from the olfactory mucosa of the nose. The vomeronasal organ is highly sensitive and detects airborne chemicals, especially pheromones, that enter the mouth.

brain where it joins the spinal cord. It is responsible for controlling the involuntary processes that keep the body alive—for example, breathing, blood pressure, and heartbeat. The hindbrain, or metencephalon, and midbrain, or mesencephalon, control basic motor functions such as balance and posture. These regions of the brain also process some sensory information and control glands that are closely associated with the brain such as the hypothalamus and pineal gland. These glands produce hormones that regulate a variety of body functions such as reproductive cycles.

The mammalian forebrain has two regions: the diencephalon, which serves as a vital relay center for sensory information; and the telencephalon, or cerebrum. The two cerebral hemispheres of the cerebrum have a creased, walnutlike surface. It is within the neurons of the cerebrum that conscious perception and control of advanced behavior such as learning and communication take place. The cerebrum of wolves includes the two enormous olfactory bulbs that receive and process information from the sensitive smelling organs in the nose. A wolf's other senses, such as vision and hearing, are also processed in centers of the cerebrum.

Sensory skills

The wolf has an extraordinarily sensitive nose. Under favorable conditions, a gray wolf can detect the scent of prey 2 miles (3.2 km) away. Scent is also important in social interactions. Each wolf has its own unique smell, and by their close association, members of a pack acquire an additional joint identity—a wolf from another pack will know the members are related.

▲ A gray wolf sniffs at the snow in search of prey. Olfaction, or smell, is the wolf's most important sense. Even in extreme weather, wolves can detect and track prey.

IN FOCUS

Seizure dogs—a sixth sense?

Guide dogs for the blind and hearing dogs for the deaf are now fairly commonplace in human society. The latest kind of assistance dog is the seizure dog. These animals are able to help people with epilepsy or other seizure-causing conditions, either by raising the alarm or by lying close to the person to stabilize him or her and help prevent serious injury. In addition, some dogs appear to be able to predict when their owner is about to have a seizure and can give a valuable few minutes' warning. Scientists are not sure how they do this, but it may have something to do with their acute sense of smell. It could be that chemical imbalances in the affected person cause him or her to give off a particular scent that the dog learns to associate with a coming seizure.

▲ Gray wolves have good vision; forward-facing eyes are a common characteristic of predators. In addition, wolves have acute hearing and a phenomenal sense of smell.

IN FOCUS

Sniffer dogs

A dog's sense of smell is directional— small changes in the concentration of scent molecules can tell the animal where a scent is coming from and which way prey was heading. This skill is put to good use by humans—trained sniffer dogs can be used to find chemical substances such as drugs or explosives, to find people lost or trapped in disaster areas, or to track missing persons. Dogs can distinguish the smell of clothes worn by different people (as long as the people are not identical twins, whose scent is identical). A trained bloodhound can follow the scent trail made by a particular person even after 24 hours, when other trails have been laid over the top.

Unlike taste, the sense of smell—also called olfaction—works over a long distance. Scent molecules are microscopic—they must be small enough to be carried in the air as vapor. An average human possesses about 5 million olfactory receptor cells in his or her nose, whereas a wolf has about 200 million. The surface area of the olfactory region inside the nose is increased by a convoluted membranous lining. If this lining were spread out flat, it would be larger than the surface area of the rest of the wolf's body.

The cells lining the olfactory region include mucus-secreting cells, pigment cells, and millions of olfactory receptor cells. These receptors trail long, hairlike cilia, which contain the scent-molecule receptors in the mucous lining of the nose. The base of each receptor cell tapers into a long narrow axon (fiberlike extension), which leads all the way to one of the two olfactory bulbs—parts of the forebrain that are located at the back of the nasal cavity.

Wolves also have excellent eyesight. The eyes face forward, with a total field of view of about 180 degrees. This range is more limited than in many prey animals, which have eyes located more to the sides of the head. But stereoscopic forward vision is advantageous to the hunter, because it aids in the perception of distance, allowing the wolf to judge leaps and pounces and move nimbly in cluttered environments, such as a forest. The peripheral vision is especially sensitive to movement—thus a wolf may spot a fleeing prey animal or another wolf out of the corner of its eye. The images gathered through the lens of the eyes are focused on the retina, on which they are sensed by specialized receptor cells called rods and cones. Rods detect monochrome light, whereas cones detect color. Wolves have good day and night vision.

Circulatory and respiratory systems

All the cells in a wolf's body require oxygen and the sugar glucose for respiration. These are delivered by the bloodstream. In vertebrates, the blood circulates in a closed system and is pumped around the body under pressure by the heart, near the center of the thorax, or chest. Like all mammals, wolves have a four-chamber heart. The left atrium (plural, atria) receives blood from the lungs. The blood then passes into the left ventricle, which pumps the oxygenated blood out through the large aorta, from which smaller arteries branch off and carry blood to the rest of the body.

Having completed a circuit, blood drains back to the heart, entering the right atrium and then the right ventricle. This ventricle pumps the blood to the lungs, where carbon dioxide is released and oxygen is absorbed. Valves located between the atria and ventricles and in the two main veins leading into the heart prevent blood from flowing in the wrong direction.

Air is drawn into the lungs through the mouth and nose. The wolf has a deep chest and large lungs. Thus it is able to breathe deeply and sustain strenuous activity such as running. However, exercise is not the only reason a wolf might breath deeply or rapidly. Because wolves do not sweat, panting is a cooling mechanism. Also, each intake of breath brings a fresh sample of air into contact with the olfactory cells in the nose and vomeronasal organ. A wolf following a trail or investigating a scent will take sharp but shallow snuffling breaths.

▼ Gray wolf
A four-chamber heart pumps blood around the body. Inhaled oxygen reaches red blood cells by way of the large lungs.

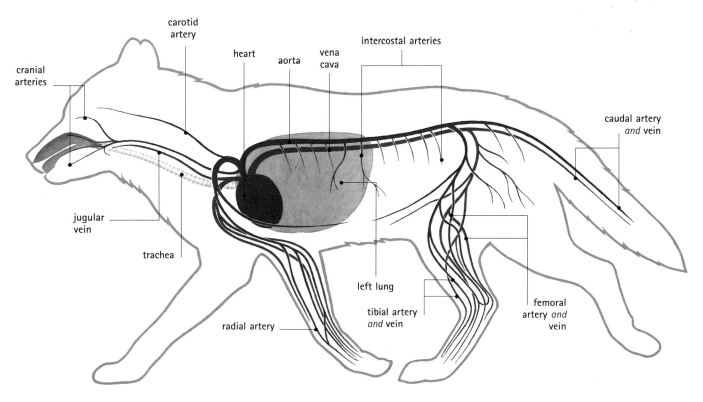

carotid artery
heart
aorta
vena cava
intercostal arteries
cranial arteries
caudal artery *and* vein
jugular vein
trachea
left lung
tibial artery *and* vein
radial artery
femoral artery *and* vein

Digestive and excretory systems

CONNECTIONS

COMPARE the teeth of a wolf with those of a *WILDEBEEST*. A wolf has canines for killing; incisors for nibbling; and carnassials for cutting meat, bone, and other tough tissues. A wildebeest is a grazer and has a reduced number of teeth. Its incisors and canines are suitable for cutting grass; and its cheek teeth are suited to grinding the grass.

Wolves have impressive teeth. They are highly modified and vary greatly in shape and function. At the front of the mouth are three pairs of incisors in each jaw. These are small and flat, with a sharp edge used for careful nibbling—for example, stripping small bits of meat off bones or grooming dirt out of the fur. On either side of the rows of incisors are single, long, pointed teeth used for stabbing and gripping prey. These lethal weapons are so characteristic of dogs that they are called canine teeth, even though most mammals have them.

Behind the canines are the cheek teeth: molars and premolars. In wolves, as in other members of the order Carnivora, the fourth upper premolars and first lower molars are modified into specialized cutting tools called carnassials. Carnassial teeth are shaped like a jagged mountain range, with two serrated main peaks, or cusps, forming a sharp cutting edge. They are very deep-rooted, to withstand the powerful forces exerted on them when the wolf gnaws on bone and other tough tissues.

Wolves do not waste much time chewing meat. Once the carnassials have sliced and diced the meat into bite-size pieces, the fragments are given a lubricating coating of saliva and are then swallowed. A wolf can consume anything up to 20 pounds (9 kg) of meat in a sitting, though the average daily consumption for a large wolf is about 5 pounds (2.5 kg).

Stomach and intestines

Digestion proper begins in the stomach, which contains a great many gland cells secreting a potent cocktail of digestive juices—stomach acid and enzymes that attack the chemical structure of the food. Meat is reduced to an acidic pulp and passes from the stomach into the first

▼ Gray wolf

The digestive process begins in the mouth, where the teeth cut up meat. Swallowed food enters the stomach, where acid and digestive juices turn the food into a pulp. As in other predatory mammals, the intestines are relatively short.

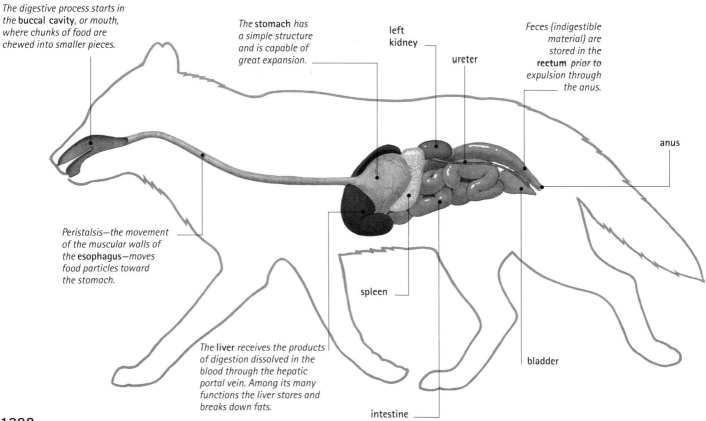

*The digestive process starts in the **buccal cavity**, or mouth, where chunks of food are chewed into smaller pieces.*

*The **stomach** has a simple structure and is capable of great expansion.*

left kidney

ureter

*Feces (indigestible material) are stored in the **rectum** prior to expulsion through the anus.*

anus

*Peristalsis—the movement of the muscular walls of the **esophagus**—moves food particles toward the stomach.*

spleen

bladder

*The **liver** receives the products of digestion dissolved in the blood through the hepatic portal vein. Among its many functions the liver stores and breaks down fats.*

intestine

IN FOCUS

A fair share

In the early stages of digestion, a wolf is able to voluntarily regurgitate food to share with other individuals. During weaning, pups are often fed this way, not only by their parents but also by other members of the pack. Even injured or elderly adults that arrive late at a kill will be offered regurgitated food by other wolves.

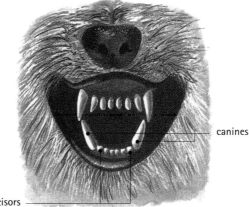

◀ TEETH
Wolves have 42 teeth. There are six incisors, two canines, eight premolars, and six molars in the lower jaw. The upper jaw has two fewer molars.

canines

incisors

section of the small intestine, the duodenum. There the pulp is blended with more digestive juices that have been secreted by the pancreas and liver. A yellowish liquid called bile, which is produced in the liver and stored in the gallbladder, is highly alkaline and helps neutralize the acids from the stomach as well as breaking down fatty molecules before the food passes farther into the intestine.

The small intestine takes one large, simple loop around the inside of the abdomen before passing into the large intestine, which is short but more convoluted. Food passes more slowly through the large intestine, allowing time for the last remnants of nutrients to be absorbed from the food. Indigestible waste passes into the rectum and out through the anus. On their way out, feces pass close to the openings of the anal glands, giving the feces a distinctive scent that is unique to each wolf. Thus other wolves can recognize the droppings that are left by animals they know.

The intestine is lined with epithelial cells. Some of these cells secrete mucus, which helps the food slide along easily, but most of the cells are concerned with absorption. The surfaces of the cells exposed to food passing along the gut have hundreds of tiny hairlike projections called villi. These greatly increase the surface area of the cell, so nutrients can be absorbed with maximum efficiency. Some of these nutrients are used by the epithelial cells themselves, but most pass through into the bloodstream for distribution to cells throughout the body. Carbohydrates are stored or converted into energy for immediate use, whereas peptides and amino acids are assembled into proteins.

Removal of metabolic wastes

The waste products of cellular metabolism are released back into the bloodstream and filtered out by the kidneys, which also remove excess water. Urine formed in the kidneys is stored in the bladder until it is expelled. Like droppings, the urine of an individual wolf has a distinctive smell and is used to mark territory.

▼ *Gray wolves are protective of their kills and will snarl at any unwelcome intruders that try to get a share.*

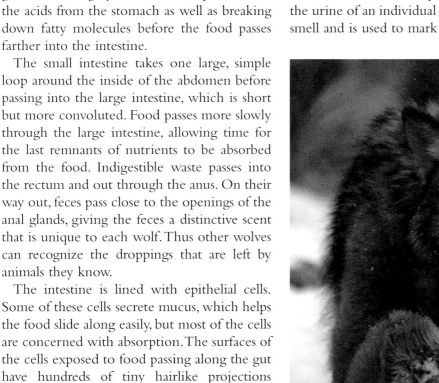

Reproductive system

Among adult wolves, males are generally about 20 percent bigger than females. Males differ anatomically in having a penis supported by a special bone, the baculum. The penis is normally tucked away inside a fold of furry skin, the prepuce. In sexually mature males, the testes hang between the back legs in a scrotal sac, which is also lightly furred. Both males and females bear eight nipples, located in pairs on the chest and belly, but only the females produce milk.

Wolves' courtship and breeding are inextricably bound up with their complex social life. The gray wolf is highly gregarious—individuals that live alone are at a serious disadvantage. Wolf packs are extended family groups and usually contain five to eight members, though sometimes several groups combine to form a large pack of 30 or more.

IN FOCUS

Stuck on you

Mating is a fairly long-drawn-out affair. The male wolf's penis, supported by the baculum bone, swells inside the female, making it virtually impossible for him to withdraw quickly. The pair usually remain locked together for at least half an hour. This looks uncomfortable, but it has a distinct advantage for the male, because as long as he remains locked in the female no other male can mate with her surreptitiously. By the time the pair disengage, the male's sperm have a head start, and there is a good chance they will fertilize the female's eggs.

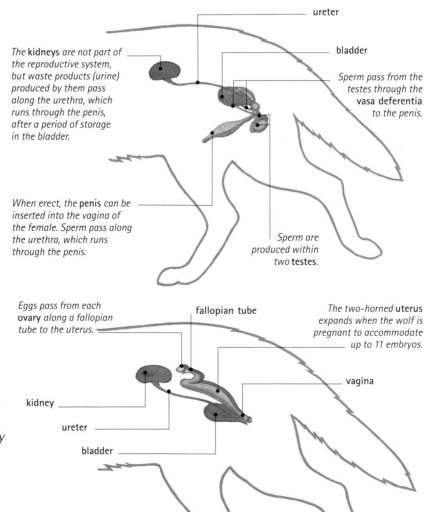

The **kidneys** are not part of the reproductive system, but waste products (urine) produced by them pass along the urethra, which runs through the penis, after a period of storage in the bladder.

ureter

bladder

Sperm pass from the testes through the **vasa deferentia** to the penis.

◄ **Male gray wolf**
Male wolves have two sperm-producing testes located in an external scrotal sac that hangs between the back legs. The penis, through which sperm is ejaculated during sex, has a supporting bone called the baculum.

When erect, the **penis** can be inserted into the vagina of the female. Sperm pass along the urethra, which runs through the penis.

Sperm are produced within two **testes**.

► **Female gray wolf**
Female wolves have two egg-producing ovaries. From each ovary, a fallopian tube leads to the uterus, which has two horns (it is bicornuate). The uterus expands greatly during pregnancy and can hold up to 11 cubs.

Eggs pass from each **ovary** along a fallopian tube to the uterus.

fallopian tube

The two-horned **uterus** expands when the wolf is pregnant to accommodate up to 11 embryos.

vagina

kidney

ureter

bladder

1400

▲ *All the adults in a pack of gray wolves take care of the cubs. By about four months old, the cubs accompany the adults on hunting excursions.*

There are separate dominance hierarchies for males and females, and usually only the dominant, or alpha, animals breed. Subordinates are usually offspring or siblings of the dominant pair. Subordinates help with the rearing of young, gaining experience that may help them become better parents themselves one day. All adults help with hunting and defend the pack territory. Though old or infirm individuals may not be much use, they are often looked after by the pack nonetheless.

Communication is vitally important in maintaining order within a pack of wolves.

GENETICS

Genes and altruism

Altruistic behavior occurs when an animal helps another animal despite the cost to itself. Wolves in a pack perform altruistic acts, such as sharing food, daily. On the face of it, this kind of behavior appears to contradict the laws of natural selection, which suggest that all animals are in competition with one another and should therefore behave selfishly. So why are wolves so good to one another? The simple answer is that all the animals in a pack are usually closely related. They share a high proportion of the same genes. Thus if an individual wolf does something to boost a relative's chances of survival and successful reproduction, it may increase the number of copies of its own genes that pass into future generations.

PREDATOR AND PREY

Lessons in life and death

Wolf cubs begin to be weaned onto regurgitated meat at five weeks, but their diet is supplemented with milk for several more weeks. Gradually they move on to more solid food, and at about four months they attempt to join in hunting. This is about the same time as they begin to shed their baby teeth. During their first hunts, cubs are clumsy and excitable and more a hindrance than a help. By seven or eight months old they have a full set of adult teeth and have gained enough experience to begin making a useful contribution to hunting excursions.

Wolves can communicate both vocally and posturally, with body language and facial expressions playing a large part. Howls and scent allow wolves to communicate over long distances, and scent messages can last a long time, so two wolves do not have to meet to exchange information.

Wolves reach sexual maturity at about two years of age, by which time they will usually have left their original pack. When a wolf has found a mate, courtship may last many weeks. Females are seasonally monoestrous—they come into breeding condition just once a year, for one to two weeks in early spring. A female may mate several times but usually with just one male, who guards her jealously. The alpha pair suppresses breeding activity in subordinate animals and constantly reassert their dominance. They will disrupt anything that resembles courtship between other wolves in the pack.

The female reproductive tract is typical for a carnivore that gives birth to a large litter. Eggs are released from the two ovaries and pass

▼ Gray wolf cubs are totally reliant on their mother's milk for up to the first five weeks of life, a period the cubs spend almost exclusively in a den. After that time, the cubs are weaned onto regurgitated meat.

▲ *An Ethiopian wolf, also called a simian jackal, regurgitates meat for her cub.*

along the ovarian ducts to the uterus. Fertilization can take place at any time after the eggs are released. The uterus is bicornuate—that is, it is a two-horn structure and can expand greatly during pregnancy to accommodate litters of up to 11 cubs, though 6 are more normal. Gestation lasts about two

◀ *During adolescence, wolf cubs play-fight, laying the foundations for their future role in the group or pack.*

months, and newborn wolf cubs weigh about 1 pound (0.45 kg). The cubs are born deaf and blind and have only a sparse covering of downy fur. Their bones are still soft (helping to avoid damage to the mother during birth), and their muscles are very weak.

To begin with, cubs move about only by crawling. They remain snuggled together for warmth in a den (usually an underground burrow) for about three weeks and are entirely reliant on milk for the first month. Each of the female wolf's nipples has about a dozen tiny pores—the openings of ducts that bring milk from the mammary glands. Lactation (milk production) is controlled by hormones and is stimulated by the cubs sucking on the nipples.

AMY-JANE BEER

FURTHER READING AND RESEARCH

Macdonald, David W. 2006. *The Encyclopedia of Mammals.* Facts On File: New York.

Macdonald, David W., and C. Sillero-Zubiri (eds.). 2004. *The Biology and Conservation of Wild Canids.* Oxford University Press: Oxford, UK.

Nowak, R. M. 1999. *Walker's Mammals of the World* (6th ed.). Johns Hopkins University Press: Baltimore, MD.

Virtual Canine Anatomy: www.cvmbs.colostate.edu/vetneuro/dissection

Woodpecker

ORDER: Piciformes FAMILY: Picidae
GENERA: *Dryocopus* and 26 others

There are more than 200 species of woodpeckers, and they live on every continent apart from Australasia and Antarctica. Woodpeckers have adapted for life in almost every type of forest habitat and many other environments besides. Most woodpeckers, but not all, are skilled at climbing vertical tree trunks. Woodpeckers live from sea level to great altitudes. For example, the Andean flicker inhabits mountain plateaus up to 16,400 feet (5,000 m) above sea level in the Andes mountains.

Anatomy and taxonomy

Scientists group all organisms into taxonomic groups based largely on anatomical features. Woodpeckers make up the family Picidae, a part of the order Piciformes. Not all members of the family are called woodpeckers: there are also piculets, flickers, sapsuckers, and wrynecks, but they share many features of anatomy and lifestyle.

- **Animals** All animals are multicellular and depend on other organisms for food. They differ from other multicellular life-forms in their ability to move around (generally using muscles) and respond rapidly to stimuli.

- **Chordates** At some time in its life cycle a chordate has a stiff, dorsal (back) supporting rod called the notochord.

- **Vertebrates** The notochord of vertebrates develops into a backbone made up of units called vertebrae. Vertebrates have a muscular system consisting primarily of muscles that are bilaterally paired (they lie on each side of an imaginary line of symmetry running along the body from head to tail). Large groups, or classes, of vertebrates include the mammals, birds, reptiles, amphibians, and fish.

- **Birds** The second largest vertebrate group, birds evolved from reptilian ancestors more than 150 million years ago. Their most obvious identifying feature is the feathers that cover the body. Birds are bipedal (walk on two legs); they do not have teeth; and at least some of their skeleton is pneumatized, or hollow. Most birds are able to fly, and all are descended from flying ancestors.

- **Order Piciformes** Scientists have traditionally placed the woodpeckers, barbets, toucans, and honeyguides (and sometimes also the puffbirds and jacamars) in the same taxonomic group because of certain shared features of their anatomy, including the tendons of their toes, their breastbone, and the palate. However, recent research on the DNA of these groups of birds has cast doubt over this scheme, so it is possible that not all the groups are particularly closely related.

- **Family Picidae** Members of this family, which is made up of woodpeckers, wrynecks, and piculets, have an anatomy that is adapted for a life climbing trees (although not all species behave in that fashion). Woodpeckers, wrynecks, and piculets have a straight or nearly straight bill, and a tongue that can be extended far beyond the end of the bill. The legs are short. Woodpeckers nest in holes, their eggs hatch after only a short period of incubation, and the young hatch naked and blind.

▼ *There are three subfamilies of woodpeckers in the family Picidae: the wrynecks, piculets, and true woodpeckers.*

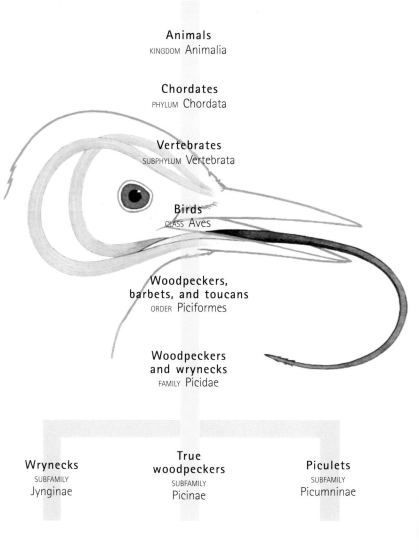

Animals
KINGDOM Animalia

Chordates
PHYLUM Chordata

Vertebrates
SUBPHYLUM Vertebrata

Birds
CLASS Aves

Woodpeckers, barbets, and toucans
ORDER Piciformes

Woodpeckers and wrynecks
FAMILY Picidae

Wrynecks
SUBFAMILY
Jynginae

True woodpeckers
SUBFAMILY
Picinae

Piculets
SUBFAMILY
Picumninae

● **Wrynecks** There are two species of wrynecks in the subfamily Jynginae. Both have cryptic (similar to their background) brown, gray, and black feathers, and the sexes are alike. The bill is relatively short and slightly curved downward, and the tongue has no barbs. Wrynecks have four toes on their feet and do not climb vertical tree trunks. Wrynecks are named for their habit of twisting their head around. They breed in natural holes and old woodpecker holes; they do not excavate their own nest hole.

● **Piculets** These tiny, short-billed woodpeckers make up the subfamily Picumninae. There are 27 species, and most live in South America. Their short tail feathers are not used as a prop as they are in larger woodpeckers. Brown colors with black markings predominate, and the sexes have different colors. The tongue is long, with fine bristles. Piculets excavate their own holes.

● **True woodpeckers** There are 182 species and 23 genera in the subfamily Picinae. The central tail feathers are strong, with sturdy shafts and vanes, well adapted for acting as a prop when the bird climbs vertically. Most species have a straight, chisel-shape bill. The nostrils are covered with bristles, and the tongue is long, with a barbed tip. True woodpeckers breed in holes (mostly in trees, but sometimes in the ground) that they dig themselves.

▲ The gila woodpecker lives in semiarid and arid regions of the southwestern United States and northern Mexico. It forages for food in treetops and on tree trunks, on cacti, and on the ground. The red patch on this bird's forehead shows it to be a male.

FEATURED SYSTEMS

EXTERNAL ANATOMY In many species strong toes and rigid central tail feathers enable woodpeckers to climb vertical tree trunks. A strong bill is used to drill through or break off bark in search of invertebrate food. *See pages 1406–1409.*

SKELETAL SYSTEM The bones are suited for flight: they are light but strong. *See pages 1410–1411.*

MUSCULAR SYSTEM The geniohyoid muscles of the head allow the tongue to be extended far beyond the tip of the bill. *See page 1412.*

NERVOUS SYSTEM Woodpeckers have acute senses of vision, hearing, taste, and touch. *See page 1413.*

CIRCULATORY AND RESPIRATORY SYSTEMS A woodpecker has a relatively large four-chamber heart, two lungs, and a system of air sacs. *See page 1414.*

DIGESTIVE AND EXCRETORY SYSTEMS Before digestion, food is broken down inside a muscular gizzard. *See pages 1415.*

REPRODUCTIVE SYSTEM Woodpeckers practice sexual reproduction. Fertilization of the female's eggs is within the single ovary. The eggs are laid in a hole in a tree or the ground, and young woodpeckers hatch from eggs. *See pages 1416–1417.*

● **Genus *Picoides*** There are about 33 species of medium-size woodpeckers in this genus, which has members in the Americas, Europe, and Asia. All are patterned black and white or brown and white, and all have a straight, chisel-tipped bill. The tail is stiff, and the first toe is usually short. Familiar North American species in the genus include the hairy woodpecker and the downy woodpecker.

External anatomy

Woodpeckers are mostly forest birds that come in a wide range of sizes and colors. At one extreme are the tiny piculets—the bar-breasted piculet of the Amazon basin is only 3 inches (7.5 cm) from bill tip to tail tip and weighs just 0.3 ounce (8 g). At the other end of the scale are very large species such as the imperial woodpecker, which was up to 24 inches (60 cm) but is probably now extinct. Between the extremes are more than 200 species of many different colors, some bright and some dull.

Varied as they are, woodpeckers have much in common. All have two wings and are capable of powered flight. Some fly very long distances. Yellow-bellied sapsuckers, for example, fly from Canada to Central America in fall and return again in spring. Woodpeckers have two legs, which most species use to climb trees; a bill, which opens into the mouth; and two eyes positioned on the sides of the head.

Green, red, black, and white
Woodpeckers have a covering of feathers over most of the body except the bill, legs, and toes. Feathers dramatically increase the surface area of the wings, making flight possible, and also provide the bird with waterproofing and insulation. Additionally, the coloration of the feathers gives each species its "branding," since no two species have exactly the same plumage.

Woodpeckers show a very wide range of colors, but some colors are more common than others: many woodpeckers are mostly green

Feathers cover the **nostrils**.

The **bill** is strong and relatively long but not as long as those of some species of woodpeckers.

The male has a crimson patch of feathers on the **nape** of the neck.

mantle

The **scapulars** and coverts lie over the flight feathers.

coverts

Two **toes** face forward and two face backward. This arrangement helps support the woodpecker when it is climbing a tree.

tertials

The **plumage** of the underparts is whitish.

There are 10 **primary feathers** in each wing. These are the longest of the flight feathers, or remiges.

The **tail** is made up of six pairs of feathers called rectrices. The stiff central pair acts as a supportive prop when the woodpecker climbs a vertical surface.

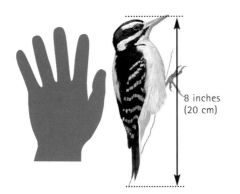

▶ **Hairy woodpecker**
The plumage is mostly black and white, but the male has a crimson mark at the back of the head. As is typical of medium-size woodpeckers, two toes face forward, and the backward-facing toes are splayed to the sides. The stiff central feathers of the tail act as a prop.

8 inches (20 cm)

▲ *The tips of a great spotted woodpecker's primary flight feathers separate when the bird flies. This action helps change the airflow over the wings—and thus the bird's movement through the air.*

with red on the head, and another large group is mostly black and white, again with red on the head. While some woodpeckers are inconspicuous in their forest habitat, others show off very bright colors. The yellow-fronted woodpecker, in South America, has bright red, bright yellow, black, and white on its head, with a blue-black back and red on the breast. Most woodpeckers are sexually dimorphic: the males and females are different colors. Usually, males have more red on the head than females, although in some species they have more yellow. Some have a pronounced crest at the back of the head; this is usually red or yellow.

The bill is a vital part of a woodpecker's anatomy, for this is its tool for finding and grasping food. It is possible to figure out a lot

IN FOCUS

Changing feathers

Woodpeckers' feathers are not permanent. They suffer a great deal of wear and tear and have to be replaced periodically. Feathers are worn out by rain, wind, and heat, and by flight and contact with the foliage of trees. Most woodpeckers replace all their flight and body feathers in late summer, and molt some of their feathers again in spring, before the breeding season begins.

Worn feathers are loosened in their sockets, or follicles, by the growth of new feathers from beneath. Eventually the old feathers are pushed out and replaced. It would be disastrous for a woodpecker if it molted all its feathers at once: it would not be able to fly and might die of cold. Such a situation does not arise, because the feathers are replaced in a sequence. The bird may be without two of its important primary flight feathers at any time during the molt period, but it is still able to fly. For example, most birds molt their tail feathers, one pair at at time, starting with the outermost. Woodpeckers that climb trees, and use their central tail feathers as a supportive prop, molt their tail in reverse: the stiffest central feathers are replaced last.

▶ *An acorn woodpecker with an acorn, for which it is named. This woodpecker collects acorns and hoards them to eat later. Note the tail feathers, which—in conjunction with the bird's toes—are used as a prop to support the bird on vertical tree trunks.*

about a woodpecker's diet by the shape of its bill. Most wood-pecking woodpeckers have a medium-length bill with a straight upper surface, a stout base, and a chisel-shape tip. This shape provides a strong structure for the birds to hammer hard at tree bark to get at the bugs living underneath. However, not all species of woodpeckers aggressively attack wood. For example, the Andean flicker has a long but very fine bill, ideal for probing in the ground for invertebrates. Piculets have a short bill—nothing like that of a typical woodpecker. These birds pick insects off the leaves of trees. A woodpecker's bill has two nostrils. Some species have bristles over the nostrils that help filter out the small particles of wood produced when they are hammering at a tree.

Toes and tails

The structure of their feet and tail allows woodpeckers varying degrees of tree-climbing skills. Most other birds have four toes at the end of each tarsus: three facing forward, and one (the hallux) facing backward. However, most woodpeckers do not have the foot structure typical of birds.

Large, relatively heavy woodpeckers have to battle gravity when they climb, more than do small species. The pileated woodpecker, for example, is able to climb vertical tree trunks only with the support of four strong toes on each leg and stiff tail feathers, which the bird uses as a supportive prop. As they climb, large woodpeckers hold their legs splayed to the side, almost as if they are trying to hug the tree.

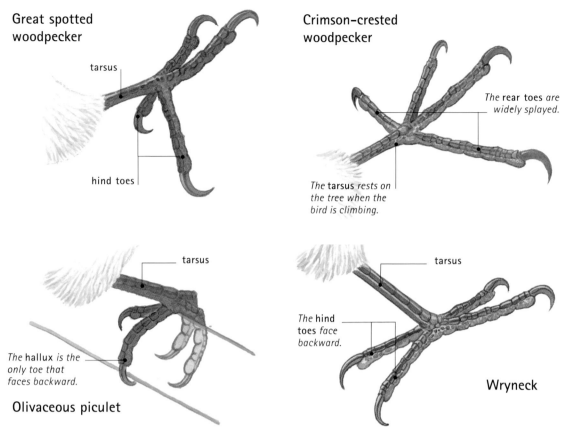

Great spotted woodpecker

tarsus

hind toes

Olivaceous piculet

The hallux is the only toe that faces backward.

Crimson-crested woodpecker

The rear toes are widely splayed.

The tarsus rests on the tree when the bird is climbing.

tarsus

The hind toes face backward.

tarsus

Wryneck

◀ TOES

In medium-size species such as the great spotted woodpecker, two toes face forward, and the two backward-facing toes are splayed. When a crimson-crested woodpecker (a large species) climbs, the hind toes are splayed out very wide, and the tarsus touches the surface.

◀ TOES

The olivaceous piculet has three toes facing forward and one back, whereas the wryneck has a zygodactyl arrangement: two toes face forward and two face backward.

All four toes point forward, and in place of the backward-pointing hallux, the "heel" at the bottom of the tarsus rests on the tree. The longish tail has six pairs of feathers, the outermost pair of which is very short. The inner tail feathers are very stiff, their shafts strengthened by longitudinal ridges. The vanes of these feathers are concave, allowing their tip to make good contact with the tree when the bird is climbing. The tail keeps the bird's body away from the tree and helps counter gravity. In addition, in flying the tail functions as a kind of aerial rudder that helps the bird change direction and acts as an air brake when the bird is preparing to land.

Smaller woodpeckers

Small and medium-size woodpeckers, such as great spotted woodpecker, are different. Some species have two toes facing forward and two facing backward; such an arrangement is called zygodactyl. Other species, such as the three-toed woodpecker, have only three toes and lack the hallux. These species also rely on a strong, stiff tail to assist them when they are climbing. Not all woodpeckers climb vertically, and

woodpeckers' toes and tail reflect this difference in behavior. Piculets are small woodpeckers that do not climb vertical tree trunks. They have three toes facing forward and one pointing back, and they do not have strengthened tail feathers. Piculets are able to grasp small branches, and they may even hang upside down from a branch when searching for invertebrate prey. However, they do not climb like other woodpeckers and so do not need to use their tail as an additional support.

IN FOCUS

Keeping dry

Many woodpeckers live in tropical rain forests where they cannot avoid getting wet. To keep their feathers from becoming waterlogged and useless, they waterproof them with oil released from the preen gland, which is situated on the rump at the base of the upper tail feathers. The oil from the preen gland also keeps the feathers flexible and inhibits the growth of bacteria and fungi. A woodpecker may rub its feathers directly over the gland or wipe its bill over the gland and then preen the feathers in need of waterproofing with its bill.

Skeletal system

The basic skeleton of a woodpecker is similar to that of other birds. Many bones that are separate in mammals are fused in birds, giving the skeleton increased strength. Also, most of the major bones are hollow, or pneumatized. The combination of fusion, strength, and lightness allows the bird to fly and to cope with the very large stresses of flight.

The vertebral column is made up of separate and fused vertebrae and runs from the skull to the pygostyle, where the muscles of the tail attach. The vertebral column has several sections. Between the skull and the first rib attachment are the cervical vertebrae, which number between 13 and 25. Next are five fused thoracic vertebrae, each of which articulates with a pair of ribs. There are seven ribs on either side of the vertebral column, but the first two pairs are tiny. The ribs attach at one end to vertebrae, and the third to seventh pairs are jointed and connect at the other end with the sternum, or breastbone. The sternum, which is linked to the shoulder area by the strong coracoid bone, varies greatly between different species of birds. The sternum has a ridge, or keel, to which the flight muscles attach. The keel is much deeper in flying birds than in flightless species. Woodpeckers are active fliers with relatively large flight muscles, and the keel of their sternum is deep.

In birds, the three lumbar, or abdominal, vertebrae are fused with the sacral vertebrae and six of the caudal, or tail, vertebrae to form

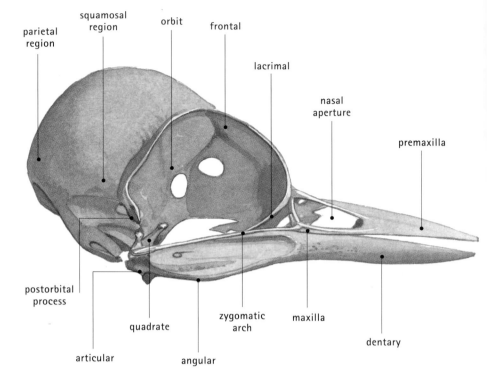

▲ SKULL
Great spotted woodpecker
The frontal, parietal, squamosal, and other bones encasing the brain are above the axis of the bill. This arrangement minimizes the effects of shock waves as the woodpecker hammers on bark.

▶ BILL
Pileated woodpecker
The pileated woodpecker has a powerful chisel-shape bill, ideally suited for hammering at tree bark. This species excavates large holes in wood in search of carpenter ants, wood-boring beetles, termites, and caterpillars.

a fused bone called the synsacrum. At the posterior (rear) end of the vertebral column is the pygostyle bone, to which the muscles of the bird's tail attach.

Bones of the wings

As in other birds, the bones of a woodpecker's two wings join the shoulder area at the glenoid fossa. From the shoulder outward, the wing bones are the humerus, radius, and ulna (together making up the "arm"); and the carpals, metacarpals, and phalanges of the digits (the "hand").

The bones of the two legs are the femur, fibula, tibiotarsus, and tarsometatarsus, and the phalanges of the digits. Each femur attaches to the ilium of the pelvis at the acetabulum. The three or four digits, or toes, of a woodpecker's feet are highly variable, with an arrangement that reflects the species' lifestyle.

Two of the other features of a woodpecker's skeleton that make it unusual are the bill and the tongue bones. The bill is a bony framework covered by a tough layer of keratin. The upper mandible of the bill is supported by the bones of the skull and is slightly mobile, unlike the upper jaw of a mammal. The bill of a hammering woodpecker has to be able to withstand great forces, and usually the shape of the bill indicates much about the lifestyle of the species. Piculets, which do not vigorously hammer at wood, have a small bill, but those species that do hammer have a chisel-shape bill, which may be long or short but is always relatively broad at the base.

◄ BILL
Olivaceous piculet
This species uses its short bill to drill for tiny holes in branches when searching for ants.

IN FOCUS

The hyoid apparatus

The tongue of a bird is supported by a collection of bones called the hyoid apparatus. The apparatus has two horns, each composed of an epibranchial and a ceratobranchial bone. The two ceratobranchial bones attach to the urohyal, basihyal, and paraglossale bones. The last is the basic tongue bone. In most birds, the hyoid apparatus is short, but woodpeckers have a long and unique apparatus that allows them to extend their tongue far beyond the end of the bill—and so reach bugs in deep crevices. The epibranchial and ceratobranchial bones, which are sheathed in the geniohyoid muscles, loop around behind the eye and meet each other again where they attach to the base of the upper mandible.

▼ HYOID APPARATUS
The bones of the tongue are called the hyoid apparatus. The central part of the apparatus is made up of three bones: the paraglossale, basihyal, and urohyal. Each of the horns of bone extending back from the basihyal is made up of two components: the ceratobranchial and the epibranchial.

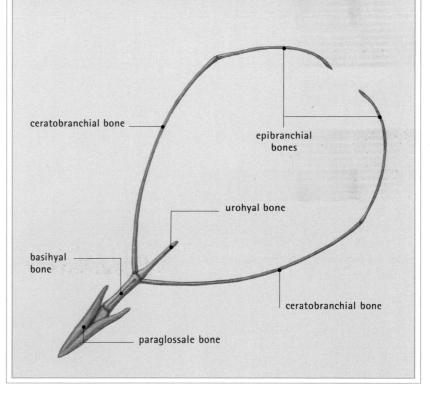

ceratobranchial bone

epibranchial bones

urohyal bone

basihyal bone

ceratobranchial bone

paraglossale bone

Muscular system

Awoodpecker has the same muscle groups as other flying birds. The muscles are dominated in size and power by the breast muscles—the pectoralis majors, which provide most of the force for flight; and the supracoracoideus muscles beneath them, which are the prime muscles responsible for raising the wings in flight. Other muscle groups in the thorax and abdomen include the external and internal oblique muscles, which hold the organs of the abdomen in place; the intercostal muscles, which strengthen the rib cage; and the serratus anterior muscles, which assist the bird in expanding and contracting its chest during respiration. Other sets of muscles move the components of the wings, the legs, the neck and head, and the tail.

Aside from the muscles of the breast, muscles are relatively large in three parts of a woodpecker's body: the neck, the legs, and the tail. Hammering woodpeckers rely on two very important bands of muscles to bend their neck quickly and strongly. The longus colli is a long band of muscle that is able to move the neck forward and down, and the semispinalis capitis (which runs along the back of the neck) flexes the neck upward and back.

Leg strength

Climbing woodpeckers have relatively powerful leg muscles, the largest of which—the gastrocnemius muscles—cover the hind surface of each leg and extend from the knee joint to the Achilles tendons to flex the digits of the feet. The levator caudae and depressor caudae enable the woodpecker to raise and lower its tail. In climbing species these are important muscles because they enable the woodpecker to use its tail as a strong prop when climbing.

CLOSE-UP

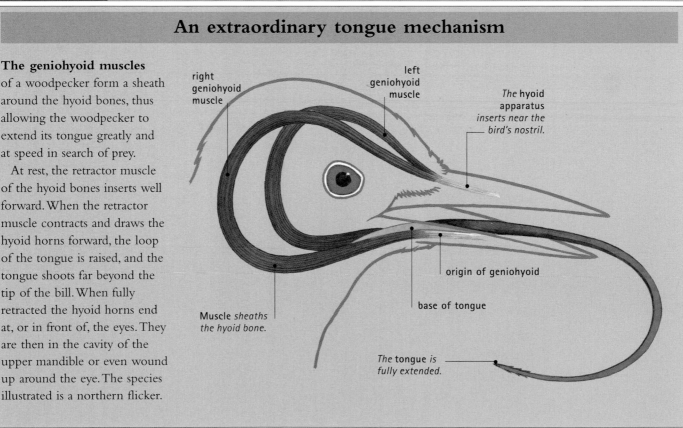

An extraordinary tongue mechanism

The geniohyoid muscles of a woodpecker form a sheath around the hyoid bones, thus allowing the woodpecker to extend its tongue greatly and at speed in search of prey.

At rest, the retractor muscle of the hyoid bones inserts well forward. When the retractor muscle contracts and draws the hyoid horns forward, the loop of the tongue is raised, and the tongue shoots far beyond the tip of the bill. When fully retracted the hyoid horns end at, or in front of, the eyes. They are then in the cavity of the upper mandible or even wound up around the eye. The species illustrated is a northern flicker.

right geniohyoid muscle

left geniohyoid muscle

The hyoid apparatus *inserts near the bird's nostril.*

origin of geniohyoid

base of tongue

Muscle *sheaths the hyoid bone.*

The tongue *is fully extended.*

Nervous system

Like all birds, woodpeckers have a brain and spinal cord (the central nervous system, or CNS) and a system of other nerves connecting all parts of the body with the brain and spinal cord (the peripheral nervous system, or PNS). The fundamental unit of the nervous system is the neuron, or nerve cell, which is able to conduct electrical impulses. The CNS receives impulses from within the body and from the environment through the sense organs and the PNS. The brain processes the information received and sends motor signals through the spinal cord and PNS to control the woodpecker's responses.

Sight, sound, taste, and touch

The senses of sight, sound, taste, and touch are very important for woodpeckers, but the sense of smell is probably less so. Sight is made possible by fibers of the optic nerves, which carry impulses from the retina of each eye to the optic fibers of the brain. Among several other roles, the glossopharyngeal nerves connect the taste buds of the tongue and the brain. The small olfactory nerves, which connect the nostrils and the olfactory lobe of the brain, provide an odor-detecting capability. Woodpeckers have a complex system of touch sensors on their tongue that identify different kinds of food.

The sense of hearing is very important for woodpeckers and other birds. They have to be able to hear the sound of an approaching predator, and woodpeckers need to hear the distant drumming of a territorial rival. Many scientists believe that woodpeckers also detect their prey partly by sound. To hear the minute movements of a bug crawling around under tree bark requires acute hearing. Like the mammalian ear, that of a woodpecker has three sections: outer, middle, and inner. The eardrum, or tympanum, at the division between the outer and middle ear chambers, vibrates when sound waves pass through the air. The vibrations pass along a tiny bone called the columnella to the cochlea of the inner ear. The cochlea translates the vibrations to impulses that are sent along the cochlear branch of the acoustic nerve to the brain, providing the sense of hearing.

IN FOCUS

How woodpeckers avoid brain damage

When a drumming woodpecker strikes hard wood it sends shock waves along the bill and into the skull. The tip of a great spotted woodpecker's bill hits tree bark at speeds up to 16 miles per hour (26 km/h). Why, then, do woodpeckers not do irreparable damage to their brain? The brain is relatively small (compared with that of many vertebrates) and is positioned above the line of impact, which runs between the tip of the bill and the quadrate bone, so the main shock waves pass below the brain. Woodpeckers have little cerebrospinal fluid around their brain; this fluid could otherwise transmit shock waves to the brain. Also, woodpeckers possess muscles that, when contracted before impact, act as shock absorbers.

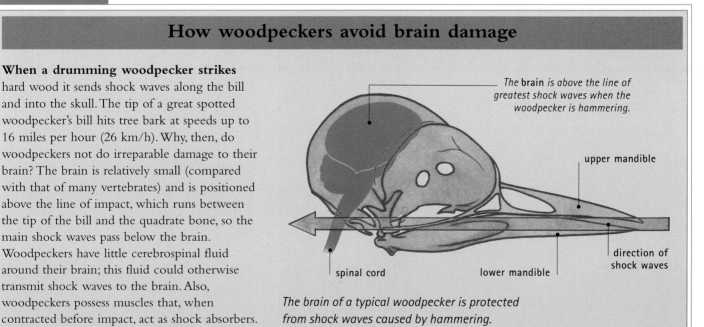

The brain is above the line of greatest shock waves when the woodpecker is hammering.

upper mandible

spinal cord   lower mandible   direction of shock waves

The brain of a typical woodpecker is protected from shock waves caused by hammering.

Circulatory and respiratory systems

▼ *This great spotted woodpecker is drumming on tree bark with its bill. Woodpeckers can make a noise either by drumming on wood with the bill or by changing the air pressure in an organ called the syrinx. The latter action vibrates a membrane that produces sound.*

Birds are very active animals, with a fast rate of metabolism, and woodpeckers are no exception. To keep the muscles supplied with the oxygen they require, and to remove the waste products of metabolism, woodpeckers' cells are supplied with a continuous supply of blood. A four-chamber heart pumps the blood around the body and keeps the circulation of oxygen-rich blood separate from the circulation of oxygen-depleted blood. In relation to body size, a woodpecker's heart is much larger than a mammal's heart, and a woodpecker's heart beats more rapidly.

Possessing a powerful heart is of no value without lungs where gas exchange can occur. Like mammals, birds have two lungs, which are connected to the mouth by a trachea and

CLOSE-UP

Making a noise

Woodpeckers are noisy birds. When proclaiming their territory, they draw attention to themselves by drumming on wood with their bill or producing loud calls from their syrinx. The syrinx is a small chamber within the bird's respiratory system. Two pairs of syringeal muscles change the volume of the syrinx. By altering the air pressure within and causing the internal tympaniform membrane to vibrate, the woodpecker can generate surprisingly loud calls.

two bronchii (singular, bronchus). Air enters the oropharynx in the rear of the mouth, and passes through the glottis valve. The valve keeps swallowed food from entering the larynx and the trachea. Unlike that of mammals, the larynx does not contain vocal cords and plays no part in the production of sound. Close to where the trachea divides into the two bronchii, which connect to the lungs, is a small chamber called the syrinx.

Air sacs

Each bronchus continues through the lung tissues to a series of thin-walled air sacs. Unlike mammals, which have to breathe out before they can inhale a fresh intake of air, woodpeckers and other birds have a series of air sacs, which increase the amount of air in the body at any one time. Gas exchange cannot take place in the air sacs, but they keep the two lungs supplied with air. Thus respiration is more efficient in birds than in mammals. Beyond the lungs, the bronchii connect to the abdominal and posterior thoracic air sacs. From the latter, air then flows forward again through the lungs and into the anterior thoracic, interclavicular, and other anterior air sacs—and then back into the trachea to be breathed out of the body.

Digestive and excretory systems

Insects and other invertebrates taken from the bark of trees, or from the ground, make up a large part of the diet of many woodpeckers. The birds reach invertebrates by hacking bark off trees with the bill, picking them off leaves, or probing for them in the ground. Other species eat berries, nuts, and fruit. American woodpeckers called sapsuckers rely mostly on tree sap for much of the year. They are able to reach this sap by hammering small holes in tree bark with their bill. Woodpeckers drink water from puddles in the forks of trees and from the ground.

Woodpeckers have a long tongue that can be extended well beyond the end of the bill. A combination of its length, barbs near the tip, and sticky saliva, which is secreted by salivary glands, enables the birds to catch insects even in deep crevices in tree trunks. Prey items and fragments of plant material are drawn into the

mouth. Since birds do not have teeth, they either swallow food whole or break it up before swallowing. Woodpeckers use several strategies to break up large food items such as acorns. If the food cannot be split by crushing between the upper and lower mandibles, some woodpeckers use an anvil to divide it. They select a fork between two tree branches and hammer the acorn into the crack. Then they hammer the food with their bill until it breaks up. Some woodpeckers have favorite anvils that they return to again and again.

The gizzard and small intestines

Food passes along the esophagus and into the proventriculus, where epithelial cells secrete strong hydrochloric acids and digestive enzymes. The chemicals start to break down the food before it enters the gizzard, a muscular chamber that plays a role similar to the teeth of mammals and turns the food into a soft pulp. Woodpeckers may regurgitate, or cough up, food from the gizzard to feed their young. Food is digested and absorbed in the small intestine, which is longer in species that eat a higher proportion of plant material than insects. Nutrients are absorbed through the walls of the intestines and pass primarily into the mesenteric veins, which bring them through the liver. There, many of the nutrients are processed and stored before passing to the body tissues. The liver stores carbohydrates and fats, creates proteins, and helps filter the waste products of metabolism in the bloodstream.

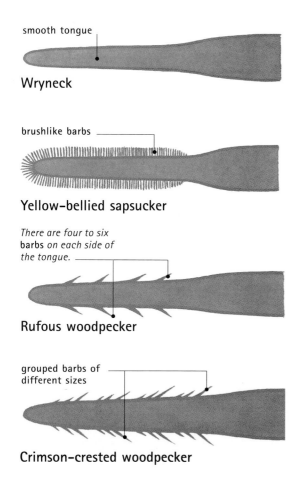

smooth tongue

Wryneck

brushlike barbs

Yellow–bellied sapsucker

There are four to six barbs on each side of the tongue.

Rufous woodpecker

grouped barbs of different sizes

Crimson-crested woodpecker

◀ TONGUES

Woodpeckers' tongues vary greatly. Species that find invertebrate prey in crevices, such as rufous woodpeckers, have a longer tongue, and long-tongue species usually have bristles concentrated near the tip. Woodpeckers that chisel deep into wood to reach prey, such as the crimson-crested woodpecker, usually have a shorter tongue with barbs. Wrynecks, which neither glean in crevices nor hammer away fragments of wood, have a relatively smooth tongue. In sapsuckers the tongue has fine brushlike bristles, ideal for collecting sap.

Reproductive system

CONNECTIONS

COMPARE the ovaries of a female woodpecker with those of an *EAGLE*. A woodpecker has only one functioning ovary (the left one), whereas a female eagle has two.

COMPARE the cloacal region of a male woodpecker with that of an *OSTRICH*. Unlike a woodpecker, an ostrich has a penis.

Like other birds, woodpeckers reproduce sexually. Male woodpeckers have two testes where male sex cells, or sperm, are produced. The testes, which are in the abdominal cavity, are adjacent to the kidneys and connect via deferent ducts to a part of the cloaca called the urodeum. For most of its length each duct is very narrow, but near the cloaca it widens into the seminal vesicle, which acts as a storage area for sperm cells. For most of the year the testes are very small organs, but during the breeding season they grow much larger. Sperm production occurs best at slightly cooler temperatures, primarily at night when the woodpecker's body temperature is slightly lower. Male woodpeckers do not have a penis, unlike some other birds.

▼ *An adult green woodpecker returns to its nest hole to feed a chick. Green woodpeckers usually have between five and eight chicks, which are fed regurgitated food by both parents.*

CLOSE-UP

Breeding hormones

In parts of the world where there are no great differences between the seasons, woodpeckers may be able to breed through most of the year. The Hispaniolan woodpecker of the Dominican Republic is an example. However, in regions where there are more clearly defined warm and cold seasons, woodpeckers mate and raise their young when conditions are warmer and food is more plentiful. Increasing hours of sunlight in the spring activate the hypothalamus, which secretes a chemical called a gonadotropin-releasing hormone, or GnRH. The GnRH activates the bird's pituitary gland, which then secretes hormones that start the production of sperm and testosterone in the testes.

COMPARATIVE ANATOMY

Birds' and reptiles' eggs

All birds and reptiles have eggs with an amniotic membrane. The membrane forms an enclosed environment that protects the developing fetus. Woodpecker eggs, like those of other birds, are both light and strong. They differ from the eggs of reptiles in important ways. The principle nutrients in a reptile's egg are proteins, but the main nutrients for the bird embryo are fats inside the yolk sac. When metabolized, the fats in a bird's egg provide more water and more energy than do the proteins in a reptile's egg. Also, the structure of a bird's eggshell allows more oxygen to enter the egg and more waste gases to escape. This ability is important, since a bird embryo has a higher metabolic rate than that of a reptile embryo.

◀ *Before they mate, black woodpeckers excavate a hole in a tree. There, the female lays three to five eggs, which are incubated by both parents for about 12 days. After the eggs hatch, the young birds are fed with food regurgitated by the parent birds.*

In a female woodpecker, unlike some birds of prey, only the left ovary and oviduct are functional. Eggs develop in the ovary, which looks like a small cluster of grapes and may contain several thousand tiny ova, or eggs. The right ovary is very small and does not play any role in reproduction. It was functional in distant ancestors of woodpeckers and is now a vestigial feature.

The cloacal kiss

When woodpeckers mate, the female crouches, and the male climbs onto her back. The male does not have a penis, so sperm passes to the female by means of a "cloacal kiss." The sperm then travels along the oviduct, and the female's eggs are usually fertilized in the infundiculum, a funnel-shape tube adjacent to the ovary. At that stage, the ovum does not look like an egg: it has no shell and no egg white. Cilia inside the infundiculum move the egg into the magnum, where a layer of egg white, or albumen, is added. This process takes about three hours. The next stage in the formation of an ovum occurs in the isthmus. There, the shell membranes are deposited on the egg, and after about an hour the egg is moved (again by the action of cilia) into the uterus, where its shell

is added. This takes longer—typically about 20 hours—and then the egg passes into the cloaca and is laid immediately.

Most birds' eggs are patterned with pigment, but woodpeckers lay white eggs. The eggs vary in shape: some species lay almost spherical eggs; others lay more pear-shaped eggs. The number of eggs also varies. A pileated woodpecker, for example, lays a clutch of two, three, or four eggs that both sexes incubate (sit on to keep warm). As with most other woodpeckers, the male pileated incubates at night, and the female sits in the day. Incubation lasts for about 18 days, and then the eggs hatch. Both parents help feed the chicks with food regurgitated from their crop. This feeding continues until a month after the chicks can fly, which occurs at age 22–26 days of age. The three-toed woodpecker is a much smaller species. The female lays three or four eggs, which hatch after 11–14 days. Both parents help feed the chicks until they can fly at 26–28 days and for up to three month after.

TIM HARRIS

FURTHER READING AND RESEARCH

Proctor, Noble S., and Patrick Lynch. 1993. *Manual of Ornithology*. Yale University Press: New Haven, CT, and London.

Winkler, Hans, David Christie, and David Nurney. 1995. *Woodpeckers: A Guide to the Woodpeckers, Piculets, and Wrynecks of the World*. Pica Press: Robertsbridge, UK.

Zebra

ORDER: Perissodactyla FAMILY: Equidae GENUS: *Equus*

The three species of zebras live in sub-Saharan Africa and have adapted to life in a variety of arid and semiarid environments. Zebras are well equipped for speeding away from predators such as lions.

Anatomy and taxonomy
Scientists group all organisms into taxonomic groups based largely on anatomical features. Zebras belong to the horse, or equid, family. Along with horses, tapirs, and rhinos, equids are part of a large group of mammals called Perissodactyla, the odd-toed ungulates. Mammals are among the most familiar of animal groups.

● **Animals** All animals are multicellular and feed off other organisms. They differ from other multicellular life-forms in their ability to move around independently (generally using muscles) and respond rapidly to stimuli.

● **Chordates** At some time in its life cycle a chordate has a stiff, dorsal (back) supporting rod called the notochord.

● **Vertebrates** In vertebrates, the notochord develops into a backbone made up of units called vertebrae. Vertebrates have a muscular system consisting primarily of bilaterally paired masses (on each side of one line of symmetry).

● **Mammals** Mammals are warm-blooded vertebrates. Fur is a unique feature of mammals, as are milk glands in the females. Also, the lower jaw is hinged directly to the skull; in this regard, too, mammals are different from all other vertebrates. Mature red blood cells in all mammals lack a nucleus; all other vertebrates have nucleated red blood cells.

● **Placental mammals** These mammals nourish their unborn young through a placenta, a temporary organ that forms in the mother's uterus during pregnancy.

● **Perissodactyls** The ungulates are a diverse group of mammals with four legs and hooves, and they are generally herbivores (plant eaters). Perissodactyls are ungulates with an odd number of toes on the hind feet at least; they have either one or three digits. The anatomical feature considered most

▼ *There are three species of living zebras, odd-toed ungulates in the genus* Equus. *The other four species in* Equus *are horses and asses. Only living species of perissodactyls are shown on this family tree.*

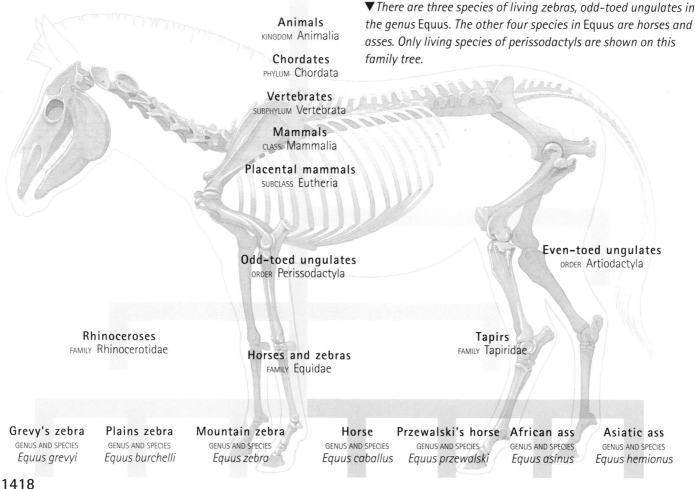

Animals
KINGDOM Animalia

Chordates
PHYLUM Chordata

Vertebrates
SUBPHYLUM Vertebrata

Mammals
CLASS Mammalia

Placental mammals
SUBCLASS Eutheria

Odd-toed ungulates
ORDER Perissodactyla

Even-toed ungulates
ORDER Artiodactyla

Rhinoceroses
FAMILY Rhinocerotidae

Horses and zebras
FAMILY Equidae

Tapirs
FAMILY Tapiridae

Grevy's zebra	Plains zebra	Mountain zebra	Horse	Przewalski's horse	African ass	Asiatic ass
GENUS AND SPECIES	GENUS AND SPECIES	GENUS AND SPECIES	GENUS AND SPECIES	GENUS AND SPECIES	GENUS AND SPECIES	GENUS AND SPECIES
Equus grevyi	*Equus burchelli*	*Equus zebra*	*Equus caballus*	*Equus przewalski*	*Equus asinus*	*Equus hemionus*

significant is that the axis of symmetry of the limbs passes through the third or middle toe. That toe is the strongest and the one on which most of the weight is borne. The 15 species of living perissodactyls are of medium or large size. They balance largely on the forelegs, and the hind legs are the main propellants. Their skeletal structure, including the firm girder of the backbone, permits fast running, and in the rhinoceroses it enables great weight to be borne. The stomach of perissodactyls is small, simple, and not divided into several chambers as in deer and antelope.

● **Rhinoceroses** The five species of living rhinoceroses are massive animals with a thick and nearly hairless hide, except in one species. They have three digits on each foot and hornlike structures on the head made of fused epidermal (skin) cells impregnated with the tough, fibrous protein keratin. The skin of rhinoceroses is very thick. The Indian and Javan rhinoceroses are covered with large, practically immovable plates, separated by joints of thinner skin to permit movement.

● **Tapirs** The four species of living tapirs are the smallest perissodactyls, along with the asses in the horse family. Tapirs are rounded, piglike, semiamphibious animals with a small proboscis (trunklike snout) and a coat of short, bristly hairs. Tapirs have primitive features, such as four hoofed toes in the forefoot and three in the hind foot, and they have relatively simple molar teeth.

● **Equids** The horses, asses, and zebras are long-legged, running perissodactyls with one functional digit in each foot. Limbs have long lower bones and digits and reduced or fused bones in the upper leg. The skull is long, with long, narrow nasal cavities. Equids have high-crowned cheek teeth for grinding plant matter.

● **Horses** The domesticated horse (*Equus caballus*) varies in appearance according to its breed. Breeds range from small

▲ *A herd of plains zebras chewing mouthfuls of grass. Each zebra's stripes are unique, with the differences being especially marked on the face and rump.*

Shetland ponies to hulking cart horses. Przewalski's horse is a descendant of the original wild horse from which domestic horses were bred.

● **Zebras** The three species of zebras are basically striped wild horses. They are easily distinguished by the pattern of stripes. The mountain zebra is the smallest species. Zebras can be further subdivided into subspecies, or local forms. The mountain zebra, for example, has two subspecies: Hartmann's zebra and the Cape mountain zebra. Subspecies are restricted to particular regions.

FEATURED SYSTEMS

EXTERNAL ANATOMY Zebras are four-legged hoofed mammals with a compact, sleek body; long legs; a flexible tail; and stripes. *See pages 1420–1423.*

SKELETAL SYSTEM A zebra's skeleton is suited to fast running, with long lower-limb bones and one very long single toe encased in a hoof. *See pages 1424–1426.*

MUSCULAR SYSTEM Muscles deliver great power to the rear legs for propulsion. *See pages 1427–1428.*

NERVOUS SYSTEM Zebras have a relatively complex brain, with an enlarged cerebellum for coordinating running with little conscious effort. *See pages 1429–1431.*

CIRCULATORY AND RESPIRATORY SYSTEMS The lungs are supplied with air through the nostrils, not from the mouth. Extra-long, dense leg arteries ensure that oxygen reaches all running muscles. *See pages 1432–1433.*

DIGESTIVE AND EXCRETORY SYSTEMS Zebras can digest a range of tough grasses and plant matter. They have a simple stomach. *See pages 1434–1435.*

REPRODUCTIVE SYSTEM A male zebra can tell when a female zebra is ready to mate by sniffing her urine. Male zebras that control female harems are more fertile than lone males. *See pages 1436–1437.*

External anatomy

CONNECTIONS

COMPARE the toes of a zebra with those of a *WILDEBEEST* and a *RHINOCEROS*. A zebra's foot has a single hoofed toe. In contrast, a wildebeest's foot has two toes with a cloven hoof, and a rhinoceros's foot has three toes, each with a hoof.

Zebras look like compact, striped horses, with their deep-chested outline, long neck, and slender legs. The mane is short and stands straight up, unlike that of most other horses, and the tail has a large tuft at the end. Apart from the stripes, those are the most obvious differences between zebras and horses. Male and female zebras look very similar, except that an adult male often has a thicker neck and is slightly bigger than an adult female.

Hairy coat

All zebras and other horses have a heavily haired coat. The zebra's coat is marked with distinctive black-and-white stripes. Differences in the patterning and width of stripes are used to distinguish among the three zebra species. The role of the stripes is not clear. Possibly,

▲ *Zebras have a flexible mouth and lips, which they use to communicate. This zebra is warning a rival.*

The **mane** *is a scrubby, erect crest of hairs extending from between the ears to the withers, or base of the neck. Stripes continue into the mane but core hairs are black.*

The **tail** *is long with horizontal black-and-white stripes ending in a tuft of long dark and pale hairs. There is a black line down the tail's center.*

The **ear**s *are flexible and taper to a point but are small in comparison with those of other zebras.*

▶ **Plains zebra**
Like other zebras, the plains zebra has a compact, sleek body and a smallish head.

49 to 53 inches (1.2–1.3 m)

83 to 102 inches (2.1–2.6 m)

Each **foot** *is made up of a single toe, in which the nail has evolved into a horny outer wall called a hoof.*

EVOLUTION

Zebra ancestors

During the early Eocene 57.8 to 52 million years ago, the first horse appeared. This hoofed, browsing mammal has been named *Hyracotherium* but is often called *Eohippus*, meaning "dawn horse." Fossils of *Hyracotherium* have been found in North America and Europe. They reveal an animal that stood just 1 to 2 feet (30–60 cm) high at the shoulder. The dawn horse had an arched back and high hindquarters. The padded feet had four hooves on the forefeet and three on the hind feet. This arrangement is unlike the unpadded, single-hoofed foot of modern horses. The teeth, skull, and brain, too, differed from those of modern horses. *Hyracotherium* was, in fact, so unhorselike that its relationship to modern equines was at first not suspected. It was only when paleontologists had unearthed fossils of more recent extinct horses that the link from *Hyracotherium* to modern horses became clear.

zebras evolved from an ancestor that lived in forests. There, stripes would have helped camouflage the animal because they disrupt its outline. In forests, stripes can look like sunshine slanting through the trees. This effect makes it harder for predators to spot striped animals.

The stripes can provide some camouflage even out on the open plain. A herd of zebras can be surprisingly difficult to see clearly where a wavy heat haze hovers close to the ground. Another interesting possibility is that stripes help protect zebras from insects that bite. Experiments have shown that tsetse flies find it harder to see striped objects than plainly marked ones. Alternatively, the stripes could be a kind of zebra uniform. Uniforms improve social cohesion, encouraging animals to stick together and look out for one another. Stripes may allow zebras to recognize other members of their herd and foals to recognize their mother. As with fingerprints, each zebra has its own unique pattern of stripes on the rump (although the general pattern conforms to that of each species), and foals quickly learn to recognize their mother's pattern.

Whatever the true origins and benefits of the zebra's striped coat, there is considerable geographical variation in the precise pattern: individuals in the extreme south of the species' range usually have fewer stripes, especially

around the rump, where the pattern sometimes fades out altogether. Some scientists believe that the extinct species from South Africa called the quagga, which had stripes only around its head, neck, and shoulders, was simply an extreme example of this trend and was thus only a subspecies of the plains zebra.

Single toes

All perissodactyls have an odd number of toes. Rhinoceroses and tapirs have three toes, but members of the family Equidae have only one. The single digit ("toe") is encased in a hoof that grows from the flesh around the foot bones. The hoof is made of keratin, as are human hair and fingernails. The zebra's hoof is thus a bit like a very thick, strong fingernail that covers a

CLOSE-UP

The hoof

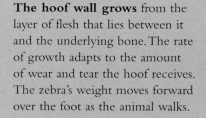

hoof wall

The hoof wall grows from the layer of flesh that lies between it and the underlying bone. The rate of growth adapts to the amount of wear and tear the hoof receives. The zebra's weight moves forward over the foot as the animal walks.

single toe. The single hoof of the equids—the only mammals to walk on the tips of single digits—is the most highly developed structure of this kind among mammals.

Like all horses, zebras have facial muscles that allow them to make a variety of expressions.

Some facial expressions are functional—for example, flaring the nostrils to smell the air—whereas others serve in communication, such as pulling back the lips to show the teeth. The ears add to the range of expressions. Usually they are pricked forward, but they can flatten

COMPARATIVE ANATOMY

Stripe patterns

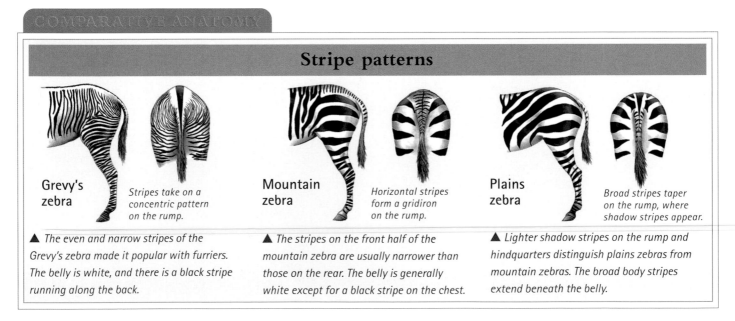

Grevy's zebra
Stripes take on a concentric pattern on the rump.

Mountain zebra
Horizontal stripes form a gridiron on the rump.

Plains zebra
Broad stripes taper on the rump, where shadow stripes appear.

▲ The even and narrow stripes of the Grevy's zebra made it popular with furriers. The belly is white, and there is a black stripe running along the back.

▲ The stripes on the front half of the mountain zebra are usually narrower than those on the rear. The belly is generally white except for a black stripe on the chest.

▲ Lighter shadow stripes on the rump and hindquarters distinguish plains zebras from mountain zebras. The broad body stripes extend beneath the belly.

Domestication and selective breeding

Horses were first domesticated 3,000 to 4,000 years ago, and the true horse *Equus caballus* now exists only in a domesticated or feral (returned to the wild) condition. Even Przewalski's horse, often referred to as the last wild horse, persists only in its native range of Mongolia, owing to an intensive conservation effort that has included the release of captive-bred individuals back to the wild. Selective breeding has resulted in several hundred breeds of domestic horses, such as elegant Arabs, thoroughbreds and quarterhorses, immensely powerful cart horses, sturdy ponies, and novelty breeds such as the tiny Falabella. The diversity of form and appearance is extraordinary within what is technically a single species.

backward to indicate aggression. A zebra's tail is long, with a tuft of long hairs starting about one-third of the way down its length. It is mobile and makes an excellent fly whisk. Zebras and other horses are social animals, and pairs often stand nose to tail alongside each other, so each animal benefits from the flicking of its partner's tail, whisking annoying insects from around the face.

▼ *These zebras have congregated to drink at a pool. Each animal has a slightly different pattern of stripes on the head. The mane is short and erect, unlike that of most domestic horses.*

▲ *There are many theories about the function of a zebra's stripes. The combined effect of a herd of stripy animals, like this plains zebra herd, might confuse predators. A lion seeing a herd might find it difficult to target an individual, especially if the herd is already on the run.*

Skeletal system

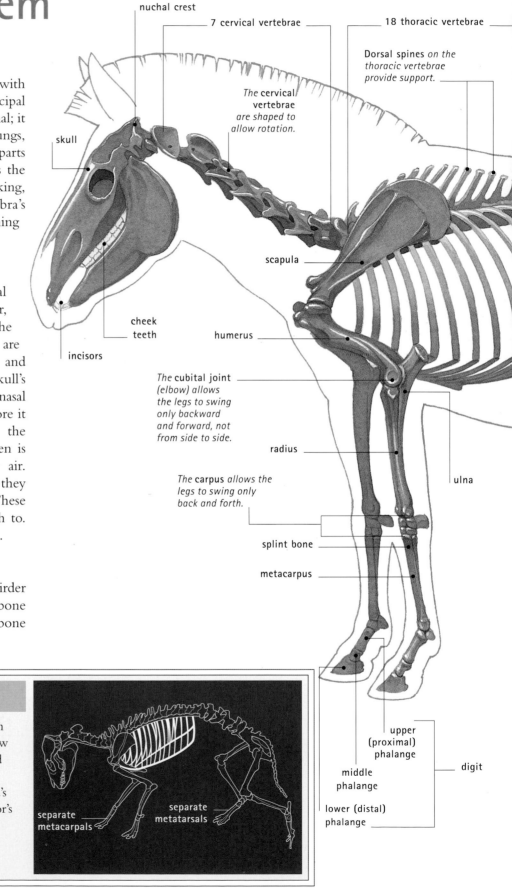

In all species of vertebrates (animals with backbones), the skeleton has four principal functions: it shapes and supports the animal; it protects vital organs such as the heart, lungs, and brain; it allows movement of body parts such as the head and feet; and it enables the animal to move from place to place by walking, running, swimming, or flying. A zebra's skeleton is particularly well suited to running at great speeds to escape predators.

Skull

In any vertebrate, the skull's principal function is to protect the brain; however, the zebra's skull is much larger than the animal's fist-sized brain requires, and there are good reasons for that. Large air passages and jawbones account for most of the skull's volume. The long and relatively broad nasal passages help ensure that air is heated before it reaches the lungs. Warm air maximizes the efficiency of the lungs, since more oxygen is absorbed from warmer air than colder air. Zebras eat lots of tough plant food, so they need strong jaw muscles for chewing. These muscles require sturdy jawbones to attach to. Large cheek teeth also require a firm base.

Backbone

The backbone, or spine, acts as a firm girder that supports the zebra's weight. The backbone is a long column made up of individual bone

EVOLUTION

The dawn horse

The skeleton of the equids' earliest known common ancestor, *Hyracotherium*, reveals how much zebras have adapted to their grassland homes, where speed is vital to escape lions and cheetahs. Most noticeable are the zebra's increase in size and longer legs. The ancestor's independent bones of the lower legs have fused in the zebras and horses, creating the metacarpus and metatarsus.

separate metacarpals

separate metatarsals

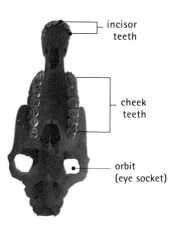

The 6 lumbar vertebrae bear stout spines and processes for muscle attachment.

The 5 sacral vertebrae are fused, giving support to the pelvic girdle.

15 to 22 coccygeal vertebrae are simple and shaped to allow movement.

sections called vertebrae. The zebra has between 51 and 58 vertebrae, extending from the skull down the back and into the tail. There are 7 neck, or cervical, vertebrae; 18 chest, or thoracic, vertebrae; and 6 lumbar, or lower-back, vertebrae. Beyond the hips, there are 5 sacral vertebrae and between 15 and 22 caudal, or coccygeal, vertebrae in the tail.

Each vertebra has a central body topped by a Y-shape arch. This structure surrounds an opening through which the spinal cord passes. The arch has a backward-pointing spine and two smaller extensions called processes. Muscles and ligaments attach to the spine and processes. The thoracic vertebrae of zebras have high dorsal spines above the front legs and ribs. These spines and the ribs support the backbone. The centrum, or central body, of each neck vertebra has a rounded head that fits into the hollow rear of the vertebra in front. This arrangement allows the zebra to look behind itself and groom its hindquarters.

Legs and limbs

The backbone balances largely on the front legs and is propelled by the hind legs. Powerful muscles that hold the

incisor teeth

cheek teeth

orbit (eye socket)

▲ JAW
Horse
The domestic horse has a jaw very similar to that of the zebra, with high-crowned cheek teeth suitable for grazing.

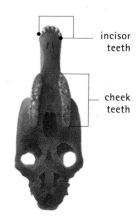

incisor teeth

cheek teeth

▲ JAW
Plains zebra
Zebras are well equipped to deal with their tough, fibrous plant diet. Strong incisors crop the plant, and the high-crowned molars, or cheek teeth, grind it efficiently.

tibia

The tarsus (hock) allows the leg to swing only backward and forward.

splint bone

metatarsus

digit

lower (distal) phalanx

CLOSE-UP

Carpus and tarsus

The carpus and the tarsus are a zebra's front "knee" and rear "knee," or hock. They evolved from the same bones that human wrist and ankle bones evolved from, so "knee" is a misleading description. The carpus is made up of eight bones, and the tarsus of six.

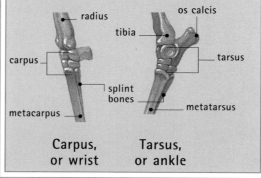

radius

os calcis

tibia

carpus

tarsus

splint bones

metacarpus

metatarsus

Carpus,
or wrist

Tarsus,
or ankle

heavy belly and provide thrust to the back legs attach to the lumbar vertebrae, which are particularly large. This arrangement of bones and muscles allows fast running.

The front and rear legs are attached to limb girdles (supports) called the scapula and the pelvic girdle respectively. The shoulder blade, or scapula, is long and narrow with a small ridge, or process, to which muscles attach. Zebras have no clavicle, or collarbone. The pelvic girdle has a broad high ilium to which the large thigh and belly muscles attach. Three other types of bones make up the pelvis: the ischium; the pubis, or pubic bone; and the fused sacral vertebrae, which provide added support. The pelvic girdle's rigid design is suited to the hindquarters' role in providing thrust; the scapula's more flexible design reflects its role as a shock absorber as well as a propulsion unit.

Over millions of years, the zebra's tiny forest-dwelling ancestor *Hyracotherium* evolved into today's long-legged running animals. In this transformation the leg bones became long and slender, and there was a loss of digits and flexibility of movement. The legs of all horses are specialized to move forward and backward, but they cannot rotate.

The increase in the length of the legs has occurred in the lower (or distal) limb bones. The humerus of the front leg and the femur of the hind leg have remained relatively short. In the foreleg, the radius and ulna have become much longer and more slender; so have the tibia and fibula in the hind leg. The joint of the short humerus with the radius and ulna permits movement only forward and backward, not from side to side.

A single toe

In the lower legs, the metacarpus and metatarsus comprise one main bone and two smaller splint bones each. These metacarpal and metatarsal bones are equivalent to the bones within human palms and insteps. A zebra's feet are made up of the bones of a single digit. The three sections, or phalanges (singular, phalanx), of this digit are very large. The hoof surrounds the distal, or lowest, phalanx. A few slivers of bones are all that remain of the other digits.

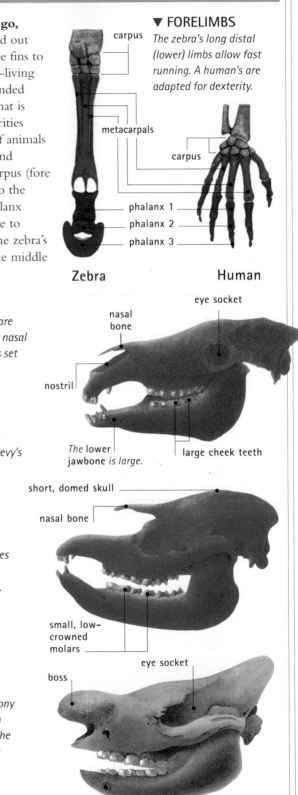

COMPARATIVE ANATOMY

Limbs and skulls

Millions of years ago, an amphibian crawled out of a swamp on leglike fins to live on land. All land-living vertebrates are descended from this creature. That is why there are similarities between the limbs of animals as diverse as zebras and humans. A zebra's carpus (fore knee) is equivalent to the human wrist. Its phalanx bones are comparable to human phalanges. The zebra's foot evolved from the middle "finger" or "toe."

▼ FORELIMBS
The zebra's long distal (lower) limbs allow fast running. A human's are adapted for dexterity.

carpus

metacarpals

carpus

phalanx 1
phalanx 2
phalanx 3

Zebra   Human

▶ SKULLS
The skulls of all zebras are long, with long, narrow nasal bones. The eye socket is set behind the teeth.

eye socket

nasal bone

nostril

The lower jawbone is large.

large cheek teeth

Plains zebra
The plains zebra's skull is somewhat larger and broader than that of Grevy's and mountain zebras. All zebra skulls have a long muzzle.

short, domed skull

nasal bone

Tapir
Short arched nasal bones and a short muzzle are typical of a tapir's skull. Tapirs have the same number of teeth as equids, but the teeth are smaller.

small, low-crowned molars

Rhino
The rhino has a large bony mound called a boss on which its horn grows. The deep sockets of the eye region allow the large neck muscles a secure anchorage.

boss

eye socket

Muscular system

CONNECTIONS

COMPARE the zebra's mobile face with that of a *BULLFROG*.

COMPARE the zebra's large leg muscles with those of the *SLOTH*.

COMPARE the zebra's powerful cheek muscles with those of a *HUMAN*.

Equids, including zebras, are built to run. The deep chest, the large lungs and heart, the length of the legs, the fusion of the limb bones, and the reduced number of toes are all adaptations for running fast; and muscle power generates the power to shift the body at dramatic speeds across open ground.

The strength and speed of equids have been exploited to great effect by humans. Relatives of the plains zebra are used all over the world as beasts of burden, as means of transport, and for sport. Racehorses are the supreme example of equids built for speed: the fastest reach about 46 miles per hour (75 km/h) for short periods.

The plains zebra, by comparison, is relatively slow, with a top speed of 34 miles per hour (55 km/h), but this is without the benefit of artificial breeding, and zebras can gallop for long periods of time if necessary.

All horses have a powerful body. The neck, chest, shoulders, and rump are particularly well muscled, whereas the limbs are slender and bony. The muscles that control the limbs are located in the shoulders and rump rather than in the legs themselves. For a human to stand or run all day would be very tiring and would require substantial leg muscles. However, even though zebras have relatively thin legs, they can remain on their feet all day without tiring. This stamina is due to the arrangement of bones and muscles in the legs, forming what is called a passive stay mechanism, which locks the knees. When the legs are locked straight, the animal's weight passes directly though the bones, which support it like the legs of a table. No muscular effort is required to keep the legs locked out.

At a full gallop, a zebra uses its hind legs to generate thrust and its forelegs to control direction and stability. Zebra muscle is dark, and rich in the pigment myoglobin, which stores oxygen for use in bouts of prolonged activity, when the amount of oxygen delivered by the bloodstream is no longer sufficient.

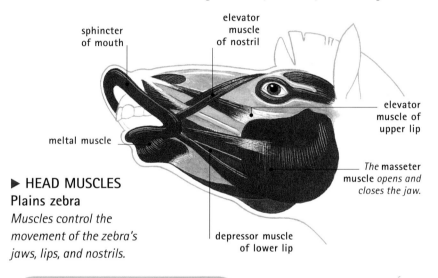

sphincter of mouth

elevator muscle of nostril

elevator muscle of upper lip

meltal muscle

The masseter muscle *opens and closes the jaw.*

depressor muscle of lower lip

▶ HEAD MUSCLES
Plains zebra
Muscles control the movement of the zebra's jaws, lips, and nostrils.

PREDATOR AND PREY

Avoiding foes

Zebra are hunted by several of large predators: lions, hyenas, leopards, cheetahs, and wild dogs readily pick off old, young, or infirm individuals, and sometimes even attack healthy adults. Zebras have a number of anatomical and behavioral adaptations that help them avoid predators. Zebras' sensitive, mobile ears; sharp, widely spaced eyes; dramatic coat patterns; and gregarious habits all help them. However, even the most alert zebra sometimes has to run for its life. At full stretch an adult plains zebra can reach 34 miles per hour (55 km/h), and its powerful leg muscles are crucial to generate the necessary rapid limb movement.

Muscle movement

triceps brachii flexed

triceps brachii relaxed

flexor muscle relaxed

flexor muscle flexed

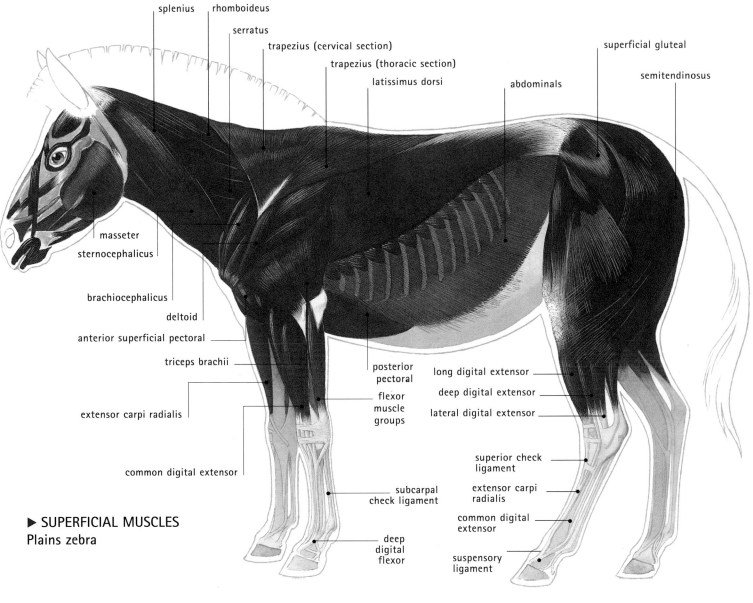

splenius
rhomboideus
serratus
trapezius (cervical section)
trapezius (thoracic section)
latissimus dorsi
superficial gluteal
semitendinosus
abdominals

masseter
sternocephalicus
brachiocephalicus
deltoid
anterior superficial pectoral
triceps brachii
extensor carpi radialis
common digital extensor

posterior pectoral
flexor muscle groups
long digital extensor
deep digital extensor
lateral digital extensor
superior check ligament
extensor carpi radialis
common digital extensor
suspensory ligament

subcarpal check ligament
deep digital flexor

► SUPERFICIAL MUSCLES
Plains zebra

▼ *When it is fleeing a predator such as a lion, a zebra uses its hind legs to generate thrust and its forelegs to control direction and stability.*

Facial muscles

The facial muscles of a zebra are dominated by large chewing muscles called masseters, which are strong enough to grind up tough vegetable matter. More delicate facial muscles control the nostrils and lips, enabling the animal to make a variety of facial expressions.

► HEAD-ON VIEW
OF SUPERFICIAL
MUSCLES
Plains zebra

trapezius
sternoephalicus
brachiocephalicus
cutaneus colli
triceps brachii
anterior superficial pectoral
extensor carpi radialis
posterior superficial pectoral

Nervous system

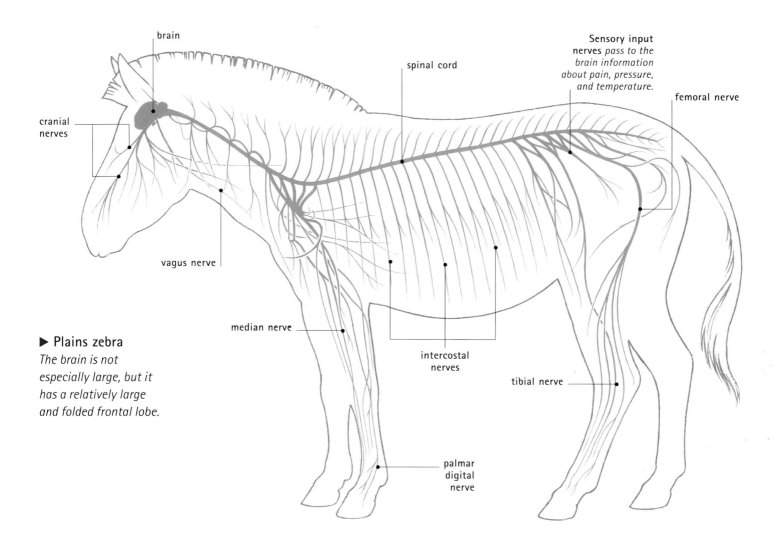

brain

cranial
nerves

spinal cord

Sensory input
nerves *pass to the
brain information
about pain, pressure,
and temperature.*

femoral nerve

vagus nerve

median nerve

intercostal
nerves

tibial nerve

palmar
digital
nerve

▶ **Plains zebra**
*The brain is not
especially large, but it
has a relatively large
and folded frontal lobe.*

CONNECTIONS

COMPARE the
position of a zebra's
eyes with the
position in an
EAGLE. The eagle
has forward-facing
eyes that make it
very good at
judging distance.
For the zebra,
peripheral vision is
more important,
because it provides
early warning of
predators.

Like all mammals, a zebra has a complex
network of nerve cells, or neurons,
connecting parts of its body. Neurons are
responsible for gathering, transmitting, and
processing sensory information and for
stimulating organs, tissues, and other body parts
to make an appropriate response. Components
of the nervous system are often referred to as
belonging to either the central nervous system
(CNS) or the peripheral nervous system (PNS).
The CNS is basically the brain and spinal cord,
and the PNS includes everything else.

The spinal cord extends from the base of the
brain along the length of the spine, where it is
protected by a sheath of tough matter and a
fatty connective tissue and by the vertebral
column itself. The cord passes directly through

IN FOCUS

Stimulus and response

All horses are alert animals, and they are
easily spooked by sudden or unexpected
events. This behavior is an adaptation for
escaping predators, when an instinctive flight
response can be the difference between life
and death. The speed of transmission of
impulses along a vertebrate's nerve averages
about 100 to 130 feet per second (30-40 m/s),
so a signal can travel to the zebra's brain and
be converted into a signal to flee within a
few hundredths of a second.

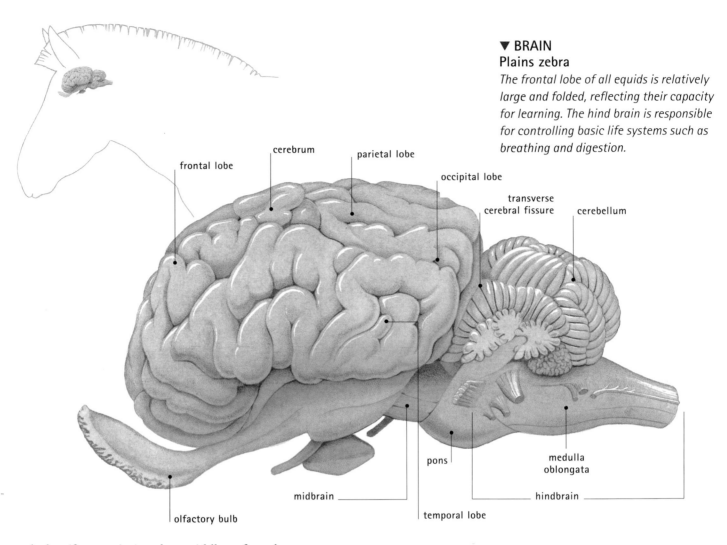

▼ BRAIN
Plains zebra
The frontal lobe of all equids is relatively large and folded, reflecting their capacity for learning. The hind brain is responsible for controlling basic life systems such as breathing and digestion.

cerebrum
frontal lobe
parietal lobe
occipital lobe
transverse cerebral fissure
cerebellum
pons
medulla oblongata
midbrain
hindbrain
temporal lobe
olfactory bulb

a hole (foramen) in the middle of each vertebra. Paired lateral (side) nerves branch off from the spinal cord at every joint between the vertebrae, and give rise to the efferent PNS. The PNS is made up mostly of motor or effector neurons that are responsible for stimulating tissues such as muscles into activity. Meanwhile, a multitude of sensory, or afferent, nerves converge at the same junctions and feed sensory information gathered all over the body back into the CNS. The branching pattern of lateral nerves is one of the few remaining clues to the segmented body plan of the simple, fishlike ancestor from which all vertebrates evolved many millions of years ago.

Functions of the brain

The equid brain is relatively well developed. It is not particularly large and occupies only the very top part of the head in the dome of the skull above the level of the eyes and extending no farther back than the back of the ears. The

IN FOCUS

Scent sensing: The flehmen response

Male zebras, horses, and several other grazing mammals sniff a female's urine to detect hormones produced when the female is in estrus (releasing eggs that can be fertilized). On the roof of the male's mouth is a pad of tissue called Jacobson's organ, or the vomeronasal organ. To detect estrus hormones, the male raises his upper lip and snorts air over this pad in an action called the flehmen response.

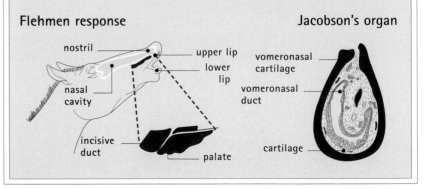

Flehmen response

nostril
upper lip
lower lip
nasal cavity
incisive duct
palate

Jacobson's organ

vomeronasal cartilage
vomeronasal duct
cartilage

brain can be divided into several regions, each with different functions. In evolutionary terms, the parts of the brain nearest the spinal cord (the hindbrain) are the oldest; they are responsible for controlling some of the most basic life-supporting functions, such as breathing and digestion. Farther forward are newer regions of the brain, which control some of the processes that set so-called "higher" animals apart from simple ones. These regions include various instinctive behaviors and some sensory processes. The forebrain, consisting of the walnut-like cerebrum, is a center for further sensory integration and higher thought processes such as memory and learning.

▼ *A noise has alerted these plains zebras, and they have turned their eyes and ears toward it. If the zebras see a predator they will turn and flee.*

IN FOCUS

Clever Hans

A hundred years ago in Germany, people were amazed by the apparent intelligence of a horse called Clever Hans. It seemed that Hans could answer arithmetic questions written on a chalkboard for him to read. Hans would respond to each question by tapping out the answer with a front foot, and he was never wrong. Scientists were astounded, and for a long time the horse had them baffled. Eventually a psychologist, Oskar Pfungst, worked out that Hans could answer questions only when the people in the room with him (in particular his trainer) could also see the board. Hans had no mathematical skills at all, but, like all horses, he was very sensitive to his surroundings. He picked up on the tiny unconscious nod given by his trainer as he approached the right answer. Hans may also have sensed other signs of tension such as an increase in the heart rates of onlookers as he got closer to a right answer. Horses are very perceptive.

Circulatory and respiratory systems

The zebra's heart is a large, powerful pump. It weighs about 10 pounds (4.5 kg) and is located low in the chest, between and just behind the front legs. The lungs are very large, filling most of the available space within the chest. The muscular diaphragm that separates the chest cavity from the abdominal cavity is also responsible for inflating the lungs. When the diaphragm contracts, the lungs expand and air is drawn in through the mouth and nose. When the diaphragm relaxes, tension in the springy tissues (the intercostal muscles and cartilage that hold the rib cage together) of the chest cavity squeezes the lungs back down and forces air back out the way it came in.

The large nostrils allow the zebra to breathe rapidly when necessary, and following prolonged exertion it is possible to see the sides of the zebra's chest pumping like bellows as the animal tries to compensate for the oxygen debt it has built up.

Arteries and veins

Oxygen-rich blood leaves the left side of the heart via a massive artery called the aorta, which curves upward and backward and runs along the top of the abdominal cavity. In a large horse, the aorta is almost the thickness of a backyard hose, with an internal diameter up to 0.4 inch (1 cm). The walls are thick, with two layers of muscle and a sheath of rubbery connective tissue. The walls are able to withstand the repeated stress of blood forced along under high pressure. Major arteries that arise directly from the aorta include the cardiac arteries supplying the heart muscle, the carotid

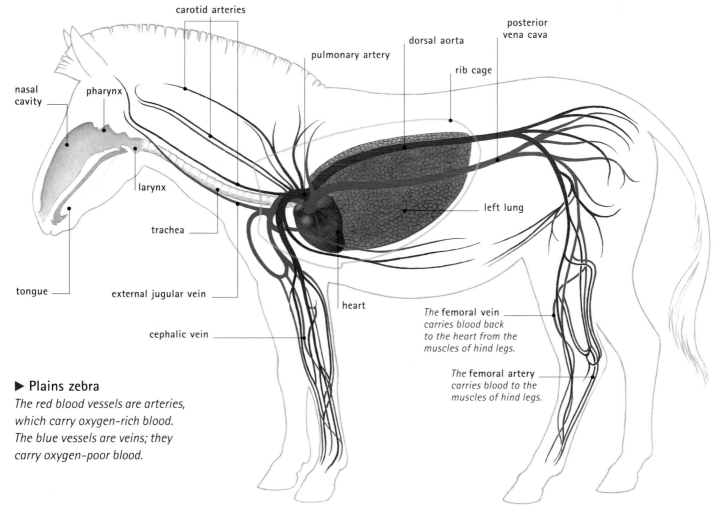

▶ Plains zebra
The red blood vessels are arteries, which carry oxygen-rich blood. The blue vessels are veins; they carry oxygen-poor blood.

Labels on diagram:
carotid arteries
pulmonary artery
dorsal aorta
posterior vena cava
rib cage
nasal cavity
pharynx
larynx
trachea
left lung
tongue
external jugular vein
heart
cephalic vein
The femoral vein carries blood back to the heart from the muscles of hind legs.
The femoral artery carries blood to the muscles of hind legs.

arteries that carry blood directly to the brain, and the celiac artery that leaves the aorta in the middle of the abdomen and then splits to form the arteries bearing the rich blood supply required by the liver, stomach, and spleen. The renal artery directs blood to the kidneys, and the femoral artery supplies the powerful muscles of the hind legs.

Alongside most of the major arteries lie large veins that drain blood from the tissues and carry it back to the heart. In highly trained horses, blood vessels can often be seen standing out from the skin, in particular on the neck and legs. The same often happens in human athletes: where the vessels lie above well developed blocks of muscle they show up because there is very little fat in the overlying skin. Blood from the right side of the heart is pumped more gently around a smaller circuit that takes in the lungs, where waste carbon dioxide is exchanged for oxygen.

IN FOCUS

Horse talk

Horses and their relatives use exhalant breaths to produce a wide range of vocalizations, including snorts, squeals, roars, and gentle huffing sounds. These sounds are more or less similar for all species of zebras. However, the sounds produced as main contact calls are distinctive. Horses use a whinny, or nickering, call, whereas asses bray. The plains zebra utters a short, harsh bark, and the mountain zebra whistles. The call of Grevy's zebra is similar to a donkey's bray and is called belling.

▼ When zebras groom one another, as these two plains zebras are doing, their heart rate falls and they become less stressed.

Digestive and excretory systems

COMPARE the digestive system of a zebra with that of a ruminant such as a *RED DEER*. The zebra has a small stomach and a large cecum, where bacteria break down cellulose, whereas in the red deer cellulose is broken down in the large multichamber stomach, and the intestine is relatively simple.

Like other horses, zebras are vegetarian. Grass forms about 90 percent of the diet, with the remainder made up of herbs and other vegetation (mainly leaves) browsed from trees and shrubs. Zebras spend more time eating than doing anything else (roughly half their life) and consume about 3 percent of their body weight in forage every day.

The processing of plant material begins the moment it is cropped from the sward. The zebra uses its broad, blade-edged incisor teeth to pluck grass and other vegetation. Each mouthful is chewed well between large, millstonelike cheek teeth. The large muscular tongue keeps the food mass churning and helps blend in saliva, which contains digestive enzymes that immediately start the process of digestion. Food is then swallowed and passes

IN FOCUS

Salts of the earth

Zebras and other horses often lick rocks and soil or even swallow chunks of earth. They do this to supplement their intake of dietary minerals. Zebras' natural diet is often lacking in certain essential minerals, in particular salt and iron, and so the minerals must be found elsewhere. Geophagia, or soil-eating, is common among zebras, and horse owners usually provide their animals with an artificial salt lick.

▼ Plains zebra
Zebras have a simple stomach and very long intestines with a saclike cecum, which houses bacteria that break down tough plant matter.

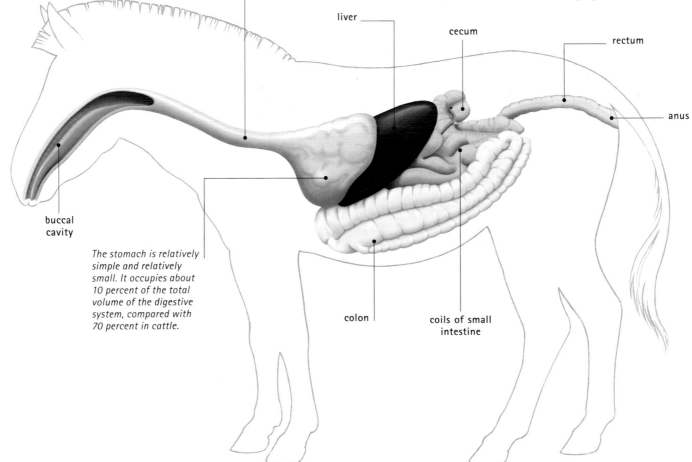

esophagus

liver

cecum

rectum

anus

buccal cavity

The stomach is relatively simple and relatively small. It occupies about 10 percent of the total volume of the digestive system, compared with 70 percent in cattle.

colon

coils of small intestine

via the esophagus to a simple stomach. The stomach is relatively small: it makes up about 10 percent of the total volume of the digestive system compared with 70 percent for the complex stomach of a cow. It is important that the zebra eats slowly but continually. Large meals cannot be accommodated in the stomach, and long periods without eating can cause the stomach to swell with gas. A zebra's intestines are extremely long, with a large saclike cecum where large numbers of bacteria aid the breakdown of plant matter such as cellulose.

A zebra's digestive system is well suited to processing large quantities of nutritionally low-grade fodder. Zebras eat long grasses with little nutritional value, but they make up for a lack of quality with quantity. These animals are able to survive on a diet that most other herbivores would find intolerable. The ability of zebras to consume very rough vegetation has important ecological effects. Areas of overgrown grassland are of little use to more refined grazers such as gazelles, whose digestive system cannot cope with rough grasses. However, once a herd of zebras has moved though an area of grassland, effectively mowing away the longer grass, other grazers can follow along behind, plucking at the fresh, tender shoots that soon begin to sprout in the zebras'

CLOSE-UP

In the horse's mouth

Adult zebras have three pairs of bladelike incisors in each jaw, used for cropping grass. The canines are small in females but large and chisel-shape in males, which use them for fighting. There are six pairs of cheek teeth (three pairs of premolars and three pairs of molars) in the lower jaw, and seven pairs in the upper jaw, which has an extra set of premolars. The cheek teeth are large, with distinctive cusps and folds in the enamel that make them very effective at grinding up plant material. The teeth appear in a predictable order in young equids. That, and an unusually consistent pattern of wear on the cheek teeth, allows zoologists to make relatively accurate estimates of a equid's age by looking inside the mouth. The expression "to look a gift horse in the mouth" is used to describe ungrateful or cynical behavior. If someone was given a horse as a gift, it would be rude to look immediately into its mouth to see if it was too old to be valuable.

wake. Plains zebras need to drink regularly and are rarely found more that 20 miles (32 km) from a water hole. Sometimes they dig for water with their front hooves, creating shallow wells. Grevy's zebras are better than plains zebras at withstanding drought and can tolerate brackish drinking water, something plains zebras and horses cannot do.

▼ *Zebras eat about 3 percent of their body weight in plant matter (mostly grass) every day.*

Reproductive system

Zebras live in herds made up of a single dominant male (stallion), a harem of mares (females), and their recent offspring. Surplus males live in smaller bachelor herds, awaiting the opportunity to set up a breeding herd of their own. Few males develop the status and experience needed to maintain a herd before the age of four years, though they are physically capable of breeding much earlier.

Female zebras reach sexual maturity when they are between 16 and 22 months old, and under ideal circumstances they are capable of producing young every year. However, because gestation lasts almost exactly 12 months, a female must mate almost immediately after giving birth, to sustain a regular annual cycle.

IN FOCUS

Precocious development

Life on the open savanna is dangerous. There is nowhere to hide, and the only real safety comes from remaining part of a herd. As an adaptation for life in these challenging circumstances, zebra foals are born in a very advanced state. The foal gets to its feet within 20 minutes of birth and can walk within an hour. Within its first few hours, it will discover the source of sustaining milk at its mother teats and attempt a skittering run. After 24 hours, it is strong and steady enough to follow its mother wherever she goes.

New mothers often come into estrus (the fertile period where females can mate and become pregnant) after giving birth. However, only those that are in exceptionally good condition will become pregnant. Females usually skip breeding for a year or even two years while rearing one youngster. Estrus lasts about a week. During this time, the soft labial tissues around the vagina swell up, and the female urinates often. The urine looks cloudy and contains pheromones, chemicals that attract the male and tell him the female will soon be ready to mate.

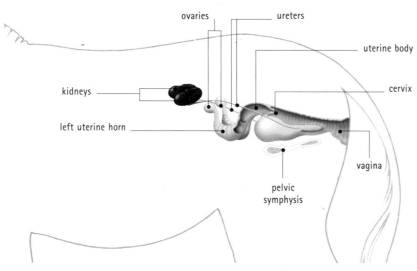

▲ FEMALE REPRODUCTIVE ORGANS
Plains zebra
Female zebras have two egg-producing ovaries and a bipartite uterus.

▶ MALE REPRODUCTIVE ORGANS
Plains zebra
Male zebras have two sperm-producing testes, suspended in the external scrotum; and a penis, through which sperm are discharged during mating.

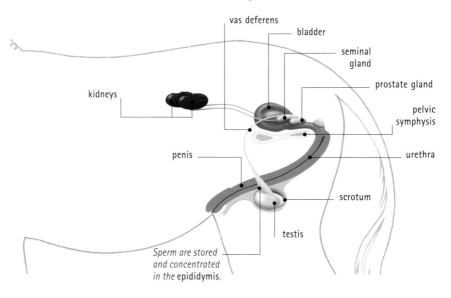

▲ *A very young zebra foal suckles from one of its mother's teats. In zebras, the bond between a mother and her newborn is very strong.*

Foals are born singly, weighing about 70 pounds (33 kg). Twins are very rare: newborn zebras are so advanced that it is virtually impossible for mares to carry two to term. Giving birth is one of the few times a female zebra will distance herself from the rest of the herd. This behavior sounds risky, but it is vital that the mother and foal spend their first few days together, away from distractions, and the mother will aggressively repel any other animal that comes too close to her foal. Young zebras are born with a powerful instinct to follow any large moving object. Usually the first thing they see is the mother, and she needs to ensure that nothing else gets in the way. The foal's life may depend on the bond it forms with its mother.

AMY-JANE BEER

FURTHER READING AND RESEARCH

Etses, R. D. 1991. *The Behavior Guide to African Mammals*. University of California Press: Berkeley, CA.

Nowak, R. 1999. *Walker's Mammals of the World* (6th ed.). Johns Hopkins Press: Baltimore, MD.

GENETICS

Hybridization

The horse family is anatomically and physiologically conservative: it exhibits nothing like the variation seen in many other mammal families such the Bovidae (sheep, cattle, antelope, and goats). The basic similarities between different species of equids permit the creation of hybrids, or crossbreeds. The most familiar of these is the mule. Mules are the result of a cross between an ass or donkey and a horse—specifically between a male ass and a female horse. The hybrid offspring of a female ass and a male horse is called a hinny. Horse–ass hybrids are always sterile because the parent species have different numbers of chromosomes. A mule receives 31 chromosomes from its father and 32 from its mother, resulting in 63 chromosomes in every mule cell. The process of meiosis, by which gametes (eggs or sperm) are produced, requires chromosomes to pair up before they replicate and divide. In a mule, the spare horse chromosome has nothing to pair up with and so the process fails every time. The same is true of various zebra–horse or zebra–ass hybrids, which are sometimes bred in captivity and are collectively called zebroids.

Index